Powders and Solids
Developments in Handling and Processing Technologies

Powders and Solids
Developments in Handling and Processing Technologies

Edited by

W. Hoyle
Consultant

RS•C
ROYAL SOCIETY OF CHEMISTRY

Chemistry Library

Special Publication No. 268

ISBN 0-85404-801-4

A catalogue record for this book is available from the British Library

Published by The Royal Society of Chemistry,
Thomas Graham House, Science Park, Milton Road,
Cambridge CB4 0WF, UK
Registered Charity Number 207890

For further information see our website at www.rsc.org

Printed and bound by Athenaeum Press Ltd, Gateshead, Tyne and Wear, UK

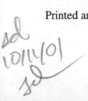

Preface

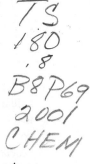

The efficient, effective, and safe handling and processing of powders is of prime importance to the chemical industry, processing as it does a high proportion of its products in powder form. In addition, the high rate of innovation means that there is the continual challenge of developing handling and processing procedures for new products.

The papers in this book present an overview of developments in the health and safety aspects of handling powders, and new technologies being applied to powders and powder handling. Measurement and control in powder handling are also covered. The papers include examples of newly introduced commercial practices, case studies and results of recent fundamental research on the behaviour of model powders.

The background to the development of the subject is covered in the first paper and illustrates why 'solids handling remains a mature industry with an immature technology'. Safety aspects are covered in the second paper and particular emphasis is placed on explosion prevention and protection measures, and the implications of new European ATEX Directives. Containment is the subject of papers three and four, the first of which describes the intermediate bulk container concept and the value of this approach for powder handling in batch processes. This is followed by a paper outlining new developments in containment of powders in the pharmaceutical industry, and in particular the selection, implementation and operation of containment strategies. More fundamental aspects of powder structure and behaviour are covered in two subsequent papers. Dodds in his paper on 'Powder Products and Structure' discusses the results of model experiments handling binary mixtures of particles of differing size ratios, and the correlation with tablet strength, dispersability, and mixing of powders. Numerical simulation of shear characteristics of granular materials is addressed by Antony et al. who report promising results using Discrete Element Modelling. Flow aid technology is another area where advances are being made. The paper 'Flow Aid Technology' reviews the range and availability of flow aids, highlights the strengths and weaknesses of present knowledge, and presents a schematic for a logical flow aid selection advisor. The last three papers cover aspects of measurement and control. In the first of the three, Campbell et al. describe current work aimed at modelling the evolution of particle size distribution throughout flour milling as a basis for the design, optimisation and control of the process. De Ryck in his paper on 'Experimental Observations on Powder Consolidation' investigates consolidation effects and kinetics for three samples of differing lumpiness using an annular shear tester. The final paper presents the successful use of electrodynamic technology in particle emission measurement and powder mass flow monitoring in a production spray dryer.

I am very grateful to all the contributors for giving so generously of their valuable time and making it possible to produce this book. I would also like to thank Gordon Lamb, David Hodge, Pat Mulqueen and David Goddard for conceiving the symposium programme which has formed the basis of these proceedings, and for their helpful comments and technical advice in preparing the text.

Bill Hoyle

Contents

Health, Safety and Handling

THE EVOLVEMENT OF SOLIDS HANDLING TECHNOLOGY

Lyn Bates
Ajax Equipment Ltd
Milton Works
Mule Street
Bolton BL2 2AR

1 INTRODUCTION

Solids Handling Technology is a multi-discipline science, with roots in civil and mechanical engineering, with stimulus from structural and chemical engineering. Its emergence as a single, comprehensive field of study followed a long gestation period in soil mechanics and a belated recognition in the bulk handling industry that particulate solids represented a fourth state of matter. The technology combines features from liquids, solids and gases, but embraces infinitely greater variation due to the interaction of a host of factors. The effect of mechanical, chemical, thermal, electrostatic and molecular features in a two or three phase media brings in further complications and the situation is made even more complex by the influence of scale, ambient conditions, industrial equipment factors and operating sequences. No wonder that the accumulation of technical knowledge took a long time to mature in order to bring a coherent structure to the subject. Any attack on a subject of this complexity must involve simplifying assumptions. Treating the media as a continuum is appropriate under certain conditions and this step allowed advances to be made.

2 HISTORY

Bulk solids have been handled and stored for thousands of years. One famous engineer from ancient times, Archimedes, calculated the number of grains of sand that would fill the then known universe as 10^{51}, displaying a basic understanding of the packing characteristics of bulk solids. His perception of density being mass divided by displacement, regardless of container shape, is also applicable to particulate solids and their variable voidage condition.

His invention of using helical screws to pump water evolved to ubiquitous applications for handling of bulk solids. Evidence of grain stores in Babylon, ancient Egypt and through the Roman Empire show that relative large volumes of bulk materials have been shipped, stored and handled from antiquity. However, it is only in recent times, with the growth of cities and the industrial revolution, that concentrations of a wide range of loose solids have been held in large gravity flow structures other than by manual handling. Whereas grain is comparatively free flowing, many other mineral and processed products display difficult flow and handling characteristics. Solids handling remains a mature industry with an immature technology.

Modern solids handling technology has its roots in soil mechanics. The stability of bridges, buildings, earthworks, dams and military fortifications has attracted the attention of engineers for centuries. The work of Coulomb[1] and Rankine[2] on friction and Reynolds,[3] who observed the dilatancy effects on sand during deformation are especially relevant to bulk solids flow. The one and only known paper of an obscure German engineer in Hamburg in 1895 was a landmark in solids handling technology, developing a theory to explain the findings of an English engineer, Isaacs Roberts,[4] of the effect of wall friction on silo wall pressures. By way of wood and glass models and simple calculus Janssen[5] developed a theory of pressure distribution in grain silos that remains today the most widely used method of assessing forces on silo walls. One might question how many papers published in this year of high technology, will be widely quoted a hundred years from now?

The maturity of centuries of bulk technology progressed at a snail pace over the first half of the next century. Verification and refinements to Janssen followed, mainly being concerned with structural aspects of silo construction. Airy,[6] Prante,[7] Toltz,[8] Ketchum,[9] Jamieson,[10] Lufft,[11] Pliessner,[12] Bovey[13] and many others, constructing experiments and calculations. Interestingly, the findings of Janssen were re-discovered by Shaxby in a joint paper with Evans,[14] during experiments with powders. Hvorslev,[15] examining the stability of cohesive soils, introduced the important concept of 'critical state' to the study of the failure characteristics of bulk material. He showed the peak stress at failure to be a function of the effective normal stress and the void ratio, and independent of the stress history of the material. The void ratio has a direct correlation with bulk density and hence provides a basis for specifying the fundamental strength potential of a compact. This concept is particularly important to an understanding of the mechanism of consolidation and flow of solids. Meanwhile, the concentration of industry, increase in the scale of production, and growing automation, highlighted the shortfall in progress in solids handling technology. The growth of mess and pollution, a thriving 'flow aids' industry and the increasing number and scale of silo failures reported by Theimer[16] all pointed to remaining deficiencies in the technology.

3 THE BREAK THROUGH

By a quirk of history a young, Polish army officer stood on a hill in 1939, with Germans advancing up one side and Russian forces up the other. It was time to pack up fighting and seek refuge in England. Andrew Jenike studied for his engineering degree in London after the war and later settled in America, working for US steel. He decided to examine a range of industrial problems and collected all the information available on about thirty subjects, collecting all the information in separate boxes that he rooted through systematically. One day he suddenly made the crucial decision that the flow of solids was one of the most important problems of the day and, being a man of very positive action, immediately scrapped all the boxes of papers collected on other topics. He approached various universities before agreeing with Utah University to research on bulk solids flow for one dollar per year.

This must have been the best scientific bargain of the century. With the aid of a young student named Jerry Johanson they developed a theory, powder property measuring instrument and design methodology that offered a solution to the age old problem of designing a silo that would guarantee the reliable discharge of non-free flowing materials. His knowledge of Russian was fortunate in that he came across a publication by a little known engineer called Sokolovskii,[17] that provided the key mathematical tool for analyzing converging flow of a plastic medium. His classical Utah Engineering Experimental Station thesis[18] published in 1964, burst upon the academic world and suddenly changed the subject of bulk solids from a black art to a respectable topic of study. The influence of charge and discharge sequences, definitive flow patterns, transient and switch stresses all fell into an understandable and predictable structure. Problems of handling loose solids, long the scourge of industry, appeared to be banished to history.

Australia instituted a combined attack on solving bulk solids storage and handling problems with the Universities of New South Wales at Sydney, Newcastle and Wollongong combining to form a centre of Bulk Solids excellence. Through the prolific words and deeds of Alan Roberts and Peter Arnold they made a strong contribution to performance of the Continent's large-scale iron and steel export industry, and spread the word internationally. Robert, Arnold and Mclean's publication[19] of a more intelligible interpretation of Jenike's work, put the theory within reach of industry.

4 THE GOLDEN YEARS

In the UK in the late 60's this science formed a part of Harold Wilson's 'White Heat of technology' that was going to transform British industry. The work of Roscoe, Schofield and Wroth on critical state soil mechanics at Cambridge University[20] fell into place. The Government set up Warren Spring Laboratories, with a major section under Dr. Fred Valentin devoted to bulk technology development. Bradford University formed a School of Powder Technology, headed Dr. John Willams. Prof. Scarlet formed a School of Particle Technology at Loughborough University. The exciting days of theory, testing and development flowed fast and furious. Devices were designed at Warren Spring Laboratories for measuring the tensile and cohesive strength of compacted bulk materials and they conducted much research into powder and paste behavior. The use of injected air for powder state control in flow was put on a scientific basis and the work broadened into investigating various forms of solids handling equipment. Walker, at the South East Electricity Generating Board developed the more user friendly Annular Shear Cell and supporting theory,[21] and Williams and Birks introduced the Unconfined Failure Tester, surely the ultimate tool for measuring arching potential.

The inter-disciplinary nature of the field of bulk solids was firmly established when Abraham Goldburg organised the first PowTech exhibition, which was quickly followed by a proliferation of similar events in UK and internationally. Specialised trade journals for the bulk solids industry were started, and more continue to be introduced. These stirrings of a coherent industrial discipline have steadily strengthened through the formation of the trade organisation SHAPA, (Solids Handling and Processing Association), which now has 100 members selling equipment to this market.

The British Materials Handling Board was set up by the government to assist the co-ordination of research into powder technology and aid the dissemination of information. Despite sterling work by its secretary Peter Middleton in organising meetings, and publishing reports and books, with limited resources it was only able to scrape the surface of the task. Relative to the scale of the industrial importance of bulk solids, the number of universities and research establishments involved throughout the world has always been surprisingly small, even though their contributions has been very significant.

5 THE STRUGGLE TO APPLY THE TECHNOLOGY

Somehow, however, British Industry never secured the extent of benefits expected. These truly golden years of promise for powder technology are not as yet fulfilled. As a 'new' subject in a consolidated form across all industries, the scale of the training problem was immense and never caught up with the backlog. Bulk technology remained a little taught subject in the syllabus of either mechanical or chemical engineering degree courses, tending to fall between the two stools of the mechanical and chemical engineering professions. This is despite the fact that over half of all the products used or consumed by humans are at some stage in a particulate form and are handled many times from source to use.

Many groups previously active have retracted or become extinct. The School of Powder technology at Bradford University was assimilated into the Chemical Engineering department with the retirement of Williams. The Loughborough School of Particle Technology never reached the same heights when Scarlet departed to Delft. The initial Government backing faded as the 'breakthrough' of efficiency was sluggish, culminating in Michael Heseltine selling the Warren Spring Laboratory site for a car park. This situation partially explains the sad reading of the Rand reports[22,23] on the lack of improvement in the solids handling industries from the 1960's onwards. The dismal record of efficiencies in plants that handle loose solids is compared with industries that handle gasses and liquids. The real reason for the lack of technical progress is that powder technology is a vast subject, greatly complicated by the host of interacting variables that influence the behavior of a mass of particulate solids.

The equipment and operator sensitivity of the apparatus, the time consuming process and the expertise needed to interpret the results of the Jenike method tended to confine application to major installations. It did not help that the theory was complex and written in a style difficult to understand. The academic world pursued even more complicated mechanical devices to reflect 'true' shear conditions, as the basic theory was considered solved, taking the subject further from widespread industrial utility. It took about ten years to develop a written procedure and provide a reference test material to give a basis for achieving consistent results, and it is still a very dubious question as to whether that objective has been achieved.[24] No wonder that the technology virtually ground to a halt, with limited shining exceptions such as at British Steel under Herbert Wilkinson and Harold Wright. The technique remains the only internationally recognised method of establishing the critical arching dimension of a bulk material in a mass flow silo to determine a 'safe' outlet size. As an example of the slow progress in the industry, the adoption of the Jenike method as a standard for ASTM in the USA, the country of its origin, has only finally gained approval during 2000.

The greatest shortcoming in the promotion of the technology was that it was in few people's interest to publicise that the value of wall friction is arguably the most important feature for design requirements. This is a simple measurement to secure. Neither was it widely promoted that static stresses are crucially important to problems of initiating flow, as emphasis on dynamic stresses for flow channel analysis tended to dominate theoretical considerations. The pragmatic requirements of industry came a poor second to the intellectual challenge of understanding the failure complexities of particulate solids in different stress conditions.

J.Schwedes, Brian Scarlet, and Geisle Enstad of TelTech in Norway, separately sought the Holy Grail of powder testing through the development of elaborate bi-axial test devices. Peichel and Schultz market versions of an annular shear tester for automatic conduction of shear tests. Michael Rotter, at Edinburgh University re-invented the Uni-axial tester for coal applications, countering the wall friction problem by double-ended compaction in the preparation cylinder by use of an elastic support for the walls. Alan Roberts in Australia and Andy Matchett of Teeside University have studied the effect of vibration on wall friction. An approach to the failure properties of cohesive powders has been made by Molerus in the light of fluid dynamics and Enstad has proposed a new theory of arching in mass flow hoppers. Chris Thornton at Aston University is pursuing the interaction of hundreds of thousands of particles in two dimensions, by means of computer simulations. The task of replicating this analysis in three dimensions is many orders of magnitude more difficult. As a spoonful of micron size particles contains in the region of 5,000,000,000 particles, and individual interactions are significantly more complicated than the simple models used, realistic replication of bulk material behavior on these lines may be expected to lie far in the future.

6 WHERE DOES THAT LEAVE US TODAY IN TERMS OF SOLVING CURRENT PROBLEMS?

Researchers are using powerful and sophisticated techniques, such as positron tracking, Discrete Element Modelling (DEM) simulation, mathematical models and finite element analysis, to examine the behaviour of loose solids. In the UK two major bulk solids centres, based respectively at Greenwich University and the Glasgow Caledonian University, provide consultation service and training for industry. Teeside, Cambridge, Surrey, Edinburgh, Birmingham, Bristol, Bath, and a few other Universities have sections active in different areas of bulk technology.

Jerry Johanson is back on the scene with a set of instruments to measure wall friction, bulk porosity and a form of shear strength. This latter device is presented with limited underlying theory and hence is not intellectually acceptable to the scientific community. His vast experience has brought him to the conclusion that the Jenike method is not appropriate to industry and a simpler approach is needed. This form of 'vertical shear cell' has been used by the author for some years as a simple tool to measure static failure conditions of a uni-axially loaded compact, a formation and failure condition reflecting incipient collapse over an opening in a non-mass flow container. An elementary comparison of the mass stimulating shear against the strength opposing failure leads to the prediction of a critical size of opening at which self weight collapse will take place. Not an elegant solution, but a conservative guide to the size of outlet needed to commence flow compared with a situation of wall slip and of re-developing flow after a dynamic stress field has been established. A wall friction device and this tool suffice to address many solids handling problems

The crying need of industry is for the degree syllabus of all chemical and mechanical engineers to include the basics of bulk technology, not exotic, ethereal techniques that are the domain of specialists. An understanding of elementary stress mechanics, friction and shear mechanisms, flow regimes, two-phase effects in compaction and dilatation and the elements of segregation processes are needed. These would enable the coming generation of equipment designers, plant and process engineers to avoid many of the pitfalls that bedevil the industry. There are also new problems, such as 'silo quaking', where large periodic forces are imposed on the structure by intermittent motion of the stored contents. Most performance shortfalls are not due to major deficiencies in the process, mechanical or structural engineering field, but arise because of apparently minor and comparatively low cost plant items do not receive the attention to function that is vital for reliable performance. The cost of equipment, such as chutes and bin outlets, has no relationship to costs of production failure.

As a matter of routine, wall friction tests should be conducted for all new solids handling and bulk storage operations where quantitative data or proven design is not available. Flow related values of shear strength can be carried out by means of simple, vertical shear tests that offer comparative values for different materials and specific measured values for incipient failure strength under defined conditions of consolidation. An appreciation of the distinction between initiating and sustaining flow and between confined and unconfined flow conditions are highly important. These should feature as crucial to criteria for defining appropriate design values for solids flow properties.

The other area of immense development interest is that of flow regime modification and other uses by means of inserts. Gravity flow generates a vertical flow promotional force, whereas the major impediment to converging flow is the reduction in the circumferential boundary of the flow channel. This is one key feature in flow modification. However, hopper inserts are fitted for many reasons, as shown in Table 1.

Table 1 *Reasons for Fitting Hopper Inserts*

Some features are independent, others inter-related, more than one objective may apply.		
Located in the Hopper Outlet Region		
- To aid the commencement of flow		
- To secure reliable flow through smaller outlets		
- To Increase solids flow rates		
- To improve the consistency of density of the discharged material		
- To secure mass flow at reduced wall inclinations		
- To expand the flow channel, secure a higher proportion of 'live' storage		
- To improve the extraction pattern		
- To prevent 'flushing'		
- To reduce overpressures on feeders		
- To save headroom/ secure extra storage capacity		
- To counter or redress segregation		
- To blend the hopper contents on discharge		
- To improve counter current gas flow distribution		
- To prevent blocking by lumps and agglomerates		
Positioned in the Body of the Hopper		
- To accelerate the de-aeration of dilated bulk material		
- To reduce overpressures/prevent threshold compaction levels		
- To alter the flow pattern for countering 'Arching' and 'Piping'		
- To Secure Mass Flow at lower wall angles		
Fixed at the Hopper Inlet		
- To secure a higher fill level	-	For personnel safety
- To reduce segregation	-	To divert oversized product
- To reduce particle attrition	-	To counter excessive wall wear

Insert technology is a valuable tool available to the designer to optimise the performance of bulk solids storage containers. The first stage of the design process is to determine a form of flow regime that is appropriate to the physical properties of the bulk material and the circumstances of the application. The shape of flow channel may be satisfied by a hopper of simple geometrical construction, such as a cone or wedge shape, but other considerations will influence the constructional method by which such a flow channel is secured. There will also remain cases where a more sophisticated approach is more viable. Inserts offer many operating advantages for a variety of applications and objectives.

The generation of mass flow in containers with relatively shallow wall angles, expanded flow, anti-arching techniques, over-pressure reduction on the contents of hoppers to reduce consolidation or high shear stresses on feeders. A vast field of opportunity is also presented in counter segregation techniques, by the use of inserts to provide multi-draw flow channels that dilute deviations from homogeneity. The ingenuity of the designer is far from exhausted in this apparently mature industrial field of solids handling. Techniques for accelerating the de-aeration of excessively dilated powders and compacting bulk materials also presents many challenges that are amenable to innovative approaches through an understanding of powder technology.

7 AND WHAT OF THE FUTURE?

The formulation of a universal data bank for bulk material properties appears superficially attractive but carries serious, if subtle hazards. There is a need for a characterisation structure, whereby the key flow related features of a new product may be quantified and the potential behavior nature identified by correlation with a substance of known manner. There is a need for wider appreciation of the importance of wall friction and different shear strength values. More text books are required on both theoretical and applied topics, to which the writer has made a small contribution[25,26] and this has been supplemented by many technical publications, some of which are available through his company.[27] Most of all, cheap, simple and user friendly test devices are needed in large quantities to provide quantified values for design and contract specifications, accompanied by undergraduate training that leads to an understanding of the key principles for their application.

References

1. C. A. Coulomb, Application des règles des maximis et minimis à quelques problèms de statique relatifs à l'architecture, *Mémoires de savants estrangers l'Académie des Science de Paris,* 1773.
2. W. Rankine, On the stability of loose earth, *Phil. Trans.,* 1857.
3. O. Renolds, On the dilitancy of media composed of rigid particles, *Phil. Mag.,* 1855, **20**, 469 – 481.
4. I. Roberts, On the pressure of wheat stored in elongated cells or bins, *Engineering,* 1882, **27 Oct.**, 399.
5. H. A. Janssen, Versuche über Getreidedruck in Silozellen. (On the Measurement of Pressures in Grain Silos), *Zeitschrift des Vereines Deutscher Ingeneieure,* 1985, 1045 – 1049.
6. W. Airy, The pressure of Grain, *Proc. of Institute of Civil Engineers,* 1897, **Vol. CXXXI**.

7. Prante, Messungen des Getreidedruckes in Silozellen, Zeitschrift des Vereines Deutscher Ingenieure, 1896, 1192.
8. M. Tolztz, *Trans. Canadian Society of Civil Engineers*, 1903, **Vol. XVII**.
9. M. S. Ketchum, The Design of Walls, Bins and Grain Elevators, Third Edition, McGraw Hill, 1919, (First published 1907).
10. J. A. Jamieson, Grain Pressures in Deep Bins, *Trans. Canadian Society of Civil Engineers*, 1903,**Vol. XVII**.
11. E. Lufft, *Engineering News*, 1904, **Vol. LII**, 531.
12. J. Pliessner, Versuche zur Ermittlung der Boden und Seitenwanddrucke in Getreidsilos, *Zeitschrift des Vereines Deutcher Ingenieure*, 1906, 976.
13. H. T. Bovey, *Trans. Canadian Society of Civil Engineers*, 1903, **Vol. XVII**.
14. J. H. Shaxby and J. C. Evans, The Variation of Pressure with Depth in Columns of Powders, *Proc. Faraday Society*, 1922, Nov., 60 – 72.
15. M. J. Hovorslev, On the Physical Properties of Distributed Cohesive Soils, *Ingeniorvidensk Skr.*, 1937, 45.
16. O. F. Theimer, Failure of Reinforced Concrete Grain Silos, *Trans. ASME Jnl. of Engng. Fir Industry, Series B*, 1969, **91, No. 2**, 460.
17. V. V. Sokolovskii, Statics of Soil Media, Butterworths, 1960.
18. A. W. Jenike, Storage and Flow of Solids, Bul. 123, The University of Utah Experimental Station, 1964.
19. P. G. Arnold, Bulk Solids Theory, Flow and Handling, The University of Newcastle Research Associates, 1982.
20. K. H. Roscoe, On the Yielding of Soils, *Geotechnique*, 1958, **8**, 22 – 53.
21. D. M. Walker, An Approximate Theory for Pressure and Arching in Hoppers, *Chem. Eng. Sci.*, 1967, **28**, 975.
22. E. W. Merrow, K. E. Phillips and C. Myers, Understanding cost growth and performance shortfalls in pioneer process plants, Rand Corp. Report, Section V, 1981.
23. E. W. Merrow, Linking R & D to Problems Experienced in Solids Processing, Summary of Rand Report, *Chem. Eng. Proc.*, 1985, (May), 14-22.
24. M. Rotter et. al., Comparison of direct shear test simulation from Britain and France, *I.Mech.E. Conf. Proc.*, 'From Powder to Bulk', June 2000, C.566/037/2000, 83 - 93.
25. L. Bates, User Guide to Segregation, British Materials Handling Board, 1997.
26. L. Bates, Guide to the Specification, Design and Use of Screw Feeders, Professional Engineering Publications for I.Mech.E., 2000.
27. Ajax Equipment Ltd., Mule Street, Bolton, BL2 2AR, UK, www.ajax.co.uk.

EXPLOSION HAZARDS IN POWDER HANDLING AND PROCESSING: THE
CHANGES AHEAD

Pieter Zeeuwen
Technical Manager - Process Safety Consultancy
Chilworth Technology Limited
Beta House, Chilworth Science Park
Southampton   SO16 7NS

1 INTRODUCTION

Most solids handled in industry are flammable and, when dispersed as a dust cloud in air, can cause a dust explosion. Experience from testing a large number of materials over many years suggests that about 70% of all dusts handled are "flammable", i.e. they are capable of exploding as a dust cloud (Figure 1).

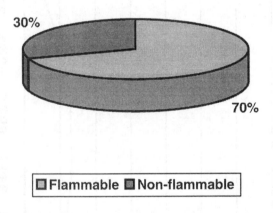

30%

70%

| Flammable | Non-flammable |

Figure 1 *70% of the dusts handled in industry are flammable*

The number of dust explosions occurring in powder handling and processing is relatively low, especially considering the number of installations with explosion hazards, showing that many installations are designed and operated with the appropriate measures in place. On the other hand, Figure 2 (based on HSE data[1]) shows that while the number of fires and explosions in general has declined over the years, the number of fires and explosions involving solids has remained virtually constant.

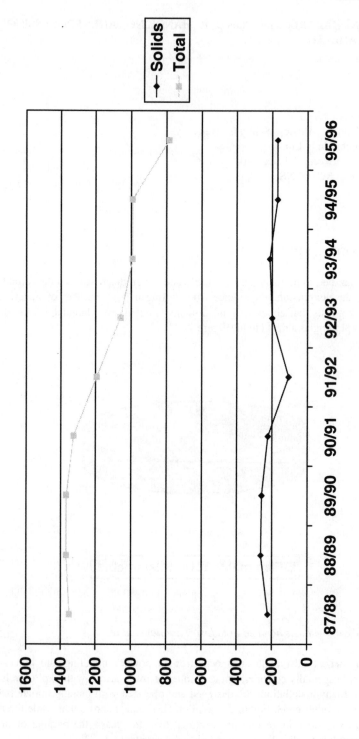

Figure 2 *Statistics for fires and explosions in the UK based on HSE data (Reference 1)*

In spite of the relatively low incidence of dust explosions, we often find that improvements would be needed to comply with statutory requirements and to operate a plant at a level of risk that is as low as reasonably practicable. Shortcomings often arise because the explosion hazards issue is considered too late in the design process, making it more difficult to implement the necessary changes. In many cases, the assumption is that adequate design and operation is possible without having proper and relevant material safety data. After an incident we often find that "assumed" material properties were not representative of the material actually handled and frequently this means that they did not err on the side of safety. It is also our experience that dust explosion issues are often considered less important than gas and vapour explosion issues. A signal of the latter is a plant where they handle flammable dusts virtually everywhere, but where the few rooms handling flammable solvents are designated the "flameproof rooms" or "Ex rooms".

Changes in Health and Safety legislation have already led to an increased requirement for risk assessments of installations. The Machine Directive requires that all machines conform to the essential safety requirements in the Directive. It is often overlooked that the hazards to consider also include explosion hazards!

Within the next few years additional European Directives will come in force that specifically address explosion hazards. One Directive, often called the "ATEX 100a" Directive, or "ATEX equipment" Directive, considers equipment (not just "machines") and protective systems for use in potentially explosive atmospheres. Gas and dust explosion hazards are both included and addressed in similar ways. All potential ignition sources (and not just electrical ones as has historically been the case) are treated equally.

The changes brought about by the ATEX 100a Directive, bringing a new system of requirements and certification for a wider range of equipment, are considerable. However, the impact of the "ATEX 137" or "ATEX work place" Directive will be greater for the operator of facilities, both new and existing.

After describing dust explosions and explosion prevention and protection measures, these new Directives and their implications will be highlighted.

2 FLAMMABLE SOLIDS

Dust explosions only occur with flammable dusts. Many solids are flammable and are recognised as such. Typical examples are fuels such as coal and wood. However, many other products such as grain, starch, milk powder, sugar, resins, polymers, metal powders, fine chemical and pharmaceutical powders, are equally flammable. Provided the particles are small enough, the majority of solids will be able to cause a dust explosion. In testing it is found that the vast majority of all dusts tested are classified as "flammable". This is true even for dusts that can hardly be ignited as a dust layer (or not at all).

For a dust explosion, one needs "dust". In this context, particles smaller than about 0.5 mm diameter are generally classified as dust. The smaller the particle size, the larger the surface area and the faster the explosion will be because the explosion reaction occurs near the particle surface (Figure 3). Unfortunately, even coarse powders usually contain at least a fraction of fine material, either because of the way they were produced or because of the handling of the material.

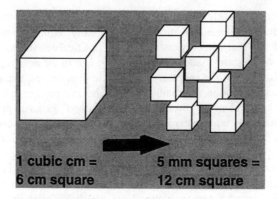

Figure 3 *Effect of particle size on the surface area of dust*

3 WHAT IS A DUST EXPLOSION?

A dust explosion resembles a gas explosion in many respects and is essentially a very rapid burning of the fuel. The difference is that the fuel is not a flammable gas but a fine flammable dust. In both cases the fuel burns very rapidly, the speed being caused by the fact that the fuel and air are mixed prior to ignition. Because the fuel and air are mixed, the explosion will consume the whole mixture once it has been ignited.

The burning of the fuel causes a large volume of hot combustion products. In an open space, this will lead to an expanding fireball. The maximum size of the fireball will be about 8 to 10 times the size of the initial dust cloud. In a closed vessel, however, expansion is not possible and the pressure will rise to about 8 - 10 times the initial pressure (see Figure 4). The time taken to reach that pressure depends on the type of dust, the dust concentration, the turbulence conditions in the cloud and the volume of the vessel. Typically, a dust explosion will take a few hundred milliseconds. This time, albeit short, provides the basis for some protective measures. The fact that the time is so short means that there is little room for correction of any mistakes.

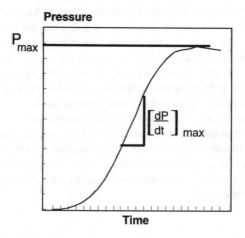

Figure 4 *Pressure-time history of a dust explosion*

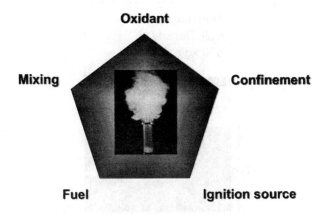

Figure 5 *Conditions for a dust explosion: the Explosion Pentagon*

The conditions for a dust explosion can be summarised as in Figure 5: one needs fuel (dust) mixed with an oxidant (usually air), (these three factors constitute a flammable mixture), an ignition source and confinement. Without confinement there will be no pressure build-up and therefore no explosion, but a flash fire.

3.1 Flame Propagation and Detonations

A situation that must be considered separately is an explosion propagating in a long duct. In this case the expansion of the combustion products causes a flow in the dust which increases the level of turbulence. Since turbulence enhances the combustion rate, the explosion will accelerate continuously until a different type of explosion propagation is reached: a detonation.

A detonation typically runs at 2 km/s (much faster than the speed of sound in the unburned mixture), has a peak pressure of about 20 bar and is more devastating and more difficult to control than a normal dust explosion. Luckily, detonations in dust-air mixtures are relatively rare - they are more common in pipelines containing a flammable gas mixture. The initial process of flame acceleration, however, is certainly very relevant because this may cause a much stronger explosion in any vessel connected to the duct.

3.2 Secondary Dust Explosions

Another feature commonly encountered in the handling of solids is the presence of dust outside the equipment. Even moderate amounts of dusts in the plant can, when raised to a dust cloud, create a flammable atmosphere. The force to raise the dust can be provided by the blast of a minor explosion in some equipment. The following, so-called secondary, dust explosion in the plant is often responsible for the collapse of whole buildings.

The quantity of dust needed for a secondary dust explosion is very small. For most dusts, a concentration of 100 g/m³ is well within the flammable range. This means that, for example, 500 grams of dust on every square meter of a floor would be sufficient to generate a flammable cloud of 5 m high, if all dust were suspended uniformly. This height would be enough to fill most plant areas. However, 500 grams of dust would only form a layer of about 1 mm thick on the floor (Figure 6). Additional dust available for secondary

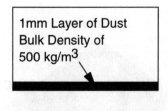

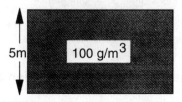

Figure 6 *Secondary dust explosions: the flammable dust cloud that can be formed from a thin dust layer on the floor*

dust explosions is often found on top of equipment, and on ridges, ledges and pipelines. This dust is often the finest and driest dust in the plant!

The issue of secondary dust explosions explains why good housekeeping must always be emphasised.

3.3   Hybrid Explosions

Hybrid explosions are explosions involving two different types of fuel: dust, gas, vapour and/or mist. In practice, the term is most commonly used for the explosion of a dust cloud where some flammable gas or vapour, often below the lower explosion limit of the gas or vapour, is present. Hybrid explosions have some special characteristics that must be taken into account in the hazard assessment and in the design of preventive and protective measures.

There are many sources for the gaseous fuel in hybrid explosions. Often it is present as part of the process, for example because solvent is used in the manufacture of the solid and the product is still solvent wet. Sometimes the solvent is still unintentionally present because it has not been removed as planned. In some cases, the gaseous fuel comes from an unrelated source – a situation that can be particularly difficult to predict.

Finally, many products release large quantities of flammable gases when they decompose. If these gases accumulate in plant, they can either be involved in a gas explosion or in a hybrid explosion.

Gas explosions and hybrid explosions will not be addressed explicitly further in this paper, but they must be considered wherever they could occur.

4 DUST EXPLOSION CHARACTERISTICS

A range of standard tests has been developed over the years in order to determine relevant characteristics of dusts. Using standard tests has the advantage that dusts can be compared and that the data can be used when applying guidelines for preventive and protective measures.

Because the dust explosion characteristics depend strongly on the particle size, shape, moisture content and contaminants, published data can only be used with great care. Besides, it is not always clear which standard has been used for the tests. This limits the applicability of the data even further as most data have been determined some time ago or in another geographical area. Consequently, in order to obtain the necessary data, tests are often necessary.

The **explosion severity** can be characterised by the maximum explosion pressure generated in a closed vessel and the maximum rate at which the pressure increases in the explosion (P_{max} and $(dP/dt)_{max}$ in Figure 4). In order to find the maximum values, a range of dust concentrations must be tested. It is common practice to consider only the maximum values. Even if it is likely that the dust concentration will not be at the optimum value, it is virtually impossible to be sure of this because deposited dust inside the equipment can always be re-suspended in the incident.

Because the maximum rate of pressure rise is volume dependent, it is common to convert that value into a volume independent parameter, K_{St}. Using the K_{St} value, dusts can be divided for convenience into so-called dust explosion classes: St 1 (includes most bulk materials and "agricultural" products), St 2 (includes many "man-made" products) and St 3 (typically metals like aluminium).

The **explosion sensitivity** of a dust can be different for different types of ignition source. Therefore different parameters must be determined to determine the sensitivity. The most common ones are the Minimum Ignition Energy (MIE) for spark ignition of a dust cloud, the Minimum Ignition Temperature (MIT) of a dust cloud (comparable to the auto-ignition temperature for gases and vapours) and the Layer Ignition Temperature (LIT) for ignition of a thin dust layer by hot surfaces.

The ignition of thick deposits or bulk material by high temperatures is subject to separate tests, where the exact conditions, especially air availability, must be tailored to represent the plant conditions as well as possible.

There are more characteristics that can be determined, like the Limiting Oxygen Concentration or the Minimum Explosible Concentration, but the need for these data must be judged case by case.

5 PREVENTING DUST EXPLOSIONS

For a dust explosion to occur, a flammable dust-air mixture is needed. Most dust handling equipment is filled with air, so the only way to prevent the flammable mixture is to avoid the dust. Depending on the type of material, a large amount of fine dust can be present at any time. Even granular material will contain some fines in most cases. During pneumatic transfer, the fines may well remain suspended long after the granules have fallen down, so that the dust concentration in the head space of the receiving vessel can be much higher than expected on the basis of the average dust content of the product.

Even if this does not happen, flammable concentrations may occur occasionally. For example, at the end of a charge, all collected fines may be fed into a silo in a short period of time. Alternatively, dust sticking to the vessel walls can be dislodged by a disturbance. A collapsing "bridge" may create flow conditions totally different from those occurring in normal operation.

So, unless the oxygen is removed from the equipment (i.e. the equipment is inerted using an inert gas instead of air), a flammable dust cloud will be present in most dust handling equipment at some point in the operation.

The other requirement for a dust explosion is the presence of an ignition source. Some 13 different types of ignition source have been identified. The most common ignition sources for dust explosions are:

- naked flames (fire)
- welding and cutting
- electrical equipment
- mechanical friction and mechanical sparks
- static electricity
- hot surfaces
- self heating, self ignition.

Naked flames (including smoking), **welding** and **cutting** are powerful potential ignition sources that must be controlled by company procedures. In some cases they can ignite a dust explosion directly, e.g. when welding on the outside of dust handling equipment. In other cases they occur indirectly because sparks or burning material enters the equipment either with the product or through openings.

Electrical equipment, when suitably selected in accordance with the hazardous area classification, and installed and maintained according to the appropriate standards, should not pose any risks. The number of ignitions by electrical equipment is consequently not very high. Unsuitable equipment or wrong use of equipment, however, still leads to incidents. Examples are the use of hand-held lamps in silos, where they can be either damaged by impact against the wall or they can be buried in the product with overheating as a consequence.

Mechanical sparks are often identified as the ignition source in a dust explosion. Further analysis of the data shows that in many cases the power of the mechanical sparks would not have been sufficient to ignite the dust, but the friction that generated the sparks will also generate a very hot surface that is definitely capable of igniting the dust cloud. In many vessels there should be no source of frictional heating or sparking. But again, often the ignition source is introduced with the product.

Static electricity occurs whenever materials are rubbed together and then separated. This means that in most solids handling, a lot of static electricity is generated. When all material is conducting and earthed, however, the static charge will not be apparent. When the product is non-conducting it will become charged even in metal plant. Similarly, even conducting product will become charged if it flows through non-conducting piping.

When the charge can accumulate and then discharge, a static ignition hazard arises. There are several forms of electrostatic discharge, ranging from a "spark" from a charged isolated conductor to "cone discharges" on the surface of bulked material in a silo. The igniting power of the various discharge types must be compared to the minimum ignition energy of the product to assess the potential hazard.

Depending on the type of powder handling, the product handled and the conditions in the plant, electrostatic ignition sources can be very important. In some cases the hazards are so intrinsic to the process, that it is impossible to prevent them and either inerting or protective measures are necessary.

Hot surfaces can arise in many ways and it is clear that they must be controlled in all places where either a dust cloud can be in contact with them or where a dust deposit can occur.

Self-heating and **self-ignition** are treacherous because they do not need any input from outside. If the material is likely to self heat in some conditions of storage temperature, moisture content and storage time, there is no way to stop it. The only prevention is to recognise the hazard and avoid the conditions altogether. Examples are reduced storage time, use of smaller silos (more heat losses) etc.

Ignition sources must always be prevented, even if the plant is protected, because the frequency of explosions must be kept as low as possible. In some cases, an additional measure can be used to reduce the frequency of ignitions: spark detection and extinguishing. Where mechanical sparks or hot particles are generated in one place (plant machinery), but the most likely location for the explosion is elsewhere (silo), it is sometimes possible to extinguish the sparks before they enter the silo. It must be noted, however, that such a system is not designed to stop an explosion; just one or more sparks.

6 EXPLOSION PROTECTION

If it is identified that an explosion could occur in plant, and that it is not possible to limit the probability to an acceptably low level, explosion protection measures are needed to control the effects of an explosion.

When considering equipment like silos, it must be recognised that the volume tends to be large. This means that any explosion will have severe consequences. This may influence the hazard assessment; if the consequences are simply unacceptable, protection is necessary even if the probability is low. On the other hand, even "small" explosions in smaller vessels have the potential of injuring or killing workers in the vicinity.

To protect plant against dust explosions, there are several options. The detailed design of protection methods is very much specialist work. Even where national or international standards are available, experience shows that the application is rarely straightforward.

The options for explosion protection are:

- explosion containment,
- explosion venting, and
- explosion suppression.

In all cases, it is necessary to prevent propagation of the explosion from one plant item to the next ("isolation"), not only to minimise the damage, but also to prevent pressure piling and flame jet ignition effects.

Containment means building the plant to be explosion resistant, i.e. so strongly that it will resist the internal pressure of the explosion. For most equipment this means a strength of around 10 bar, but the actual design pressure must be determined taking into account all relevant parameters (initial pressure, temperature, etc.). Obviously, such a high design pressure is rarely an option for large silos, but for smaller hoppers it may well be feasible. Because the full explosion pressure is contained, the maximum force is generated for propagation to connected equipment, so isolation is of the utmost importance.

Two options have been developed over the years: explosion pressure resistant and explosion pressure *shock* resistant. The latter makes better use of the strength of the

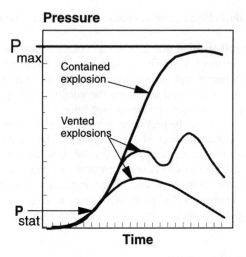

Figure 7 *Pressure-time history during a vented dust explosion. P_{stat} is the opening pressure of the venting device, P_{max} the maximum pressure that would be obtained in a closed vessel without explosion vents fitted*

construction. This is deemed acceptable because an explosion is a one-off event of short duration. Obviously, the probability that some deformation of the equipment occurs is increased.

Explosion venting is sometimes called the "natural" explosion protection. This is because without any protection, the equipment would fail and the explosion would be vented. Because of the venting, the maximum pressure during the explosion would be reduced (see Figure 7). To apply explosion venting in a controlled manner, however, is not as simple as this analogy suggests. It is imperative that a vent opening of sufficient size is available at the right time. Because of the short duration of the explosion, there is no time to overcome the inertia of heavy covers. In many explosions where "vent covers" were designed in the wrong manner, the explosion pressure increased so fast that the vessel ruptured before the vent was even opened!

For large silos the necessary vent area is large and there may be problems to accommodate the vent on the silo roof, especially if the roof is not free of obstructions. Many silos are elongated and this must be taken into account in the design of the vent area; in the extreme a silo can resemble a duct, and what that means for the explosion process inside has been explained before.

Venting is in reality nothing more than displacing the explosion to another location; when the vent opens, both unburned and burnt dust is expelled from the vent and this creates a huge flame jet or fireball outside the vent opening. The size of this flame is often underestimated and the safe area is either too small or lacking altogether. Also the pressure is vented from the vessel and that can cause some damage to the surroundings.

Because of the external effects, any equipment located indoors can only be vented safely via a duct to a safe location outside. The vent duct, however, will reduce the efficiency of the venting and either the vent must be significantly enlarged or the vessel

Figure 8 *Explosion test in a bucket elevator, showing how the explosion propagates the length of the elevator in spite of the explosion vents (photo courtesy of the Health & Safety Laboratories)*

must be built stronger. For smaller vessels nowadays some special "flameless venting devices" are available that allow venting indoors.

Explosion venting does not stop the explosion. Therefore, even if a vent is present close to the ignition location, the explosion will propagate throughout the equipment. This can be seen in Figure 8, taken during a test at the Health & Safety Laboratories in Buxton: with explosion vents located along the length of this bucket elevator, the explosion still propagates the length of the elevator. This explosion behaviour has implications for elongated equipment such as bucket elevators, where it may be difficult to find suitable safe areas especially near the lower parts of the elevator.

Explosion suppression relies on the detection of the incipient explosion and the injection of a suitable suppressant from one or more suppressors (see Figure 9). This effectively extinguishes the explosion and so limits the pressure rise to a value that the vessel can withstand. Suppression systems must be designed to work effectively in the time available and the design is therefore dependent on the hardware of the particular supplier. Especially on large volumes like silos, the investment can be considerable. Suppression of large silos is therefore very much limited to those cases where venting is not permissible (e.g. for environmental reasons), and inerting (as a preventive measure) or containment are not feasible. For smaller vessels the range of application is much wider.

Explosion isolation is, as mentioned before, needed to prevent propagation of an explosion from one vessel to the next. This is not just necessary to limit the damage, but

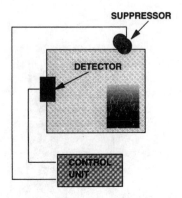

Figure 9 *Schematic of an explosion suppression system fitted to a vessel*

also because explosions become more difficult to protect against as they propagate. Two effects are important here: flame acceleration and pressure piling. Consider the situation in Figure 10 of two interconnected vessels. When the explosion starts in one vessel, the pressure will increase in the other vessel before the ignition propagates into the second vessel. Thus the explosion will occur at an elevated initial pressure, and the explosion pressure will be proportionally higher. If the pressure in the second vessel is only 1 bar g at the time of ignition, the maximum pressure in this vessel will already be about 19 bar instead of 9 bar (10 times the absolute initial pressure)!

Additionally, the flow of the gases from the primary vessel to the secondary vessel will increase the level of turbulence there (which will increase the severity of the explosion as mentioned earlier) and the ignition source will be a large flame jet instead of the usual point source.

A range of measures is available that can be implemented. The best, in terms of value for money, are those that do not need additional investment. The first of this kind is to simply avoid any connections between vessels that are not absolutely necessary. The second is the use of an explosion proof rotary valve instead of a standard one. Such rotary valves are stronger than normal and have smaller gaps (which stop the flame). Rotary valves need to be stopped in case of an explosion, because otherwise burning material may be transferred to the next vessel.

Other options are the use of special valves to stop the explosion in a duct, a diverter (local venting arrangements on a duct) and advance inerting barriers. Each option has its advantages and disadvantages.

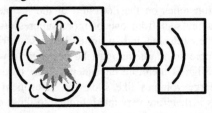

Figure 10 *Pressure piling in interconnected vessels: the pressure form the explosion in the primary vessel (left) will increase the pressure at which the explosion starts in the secondary vessel (right)*

7 THE ATEX DIRECTIVES

The above basics about dust explosions and explosion prevention and protection have been well established for many years. What is changing at the moment is that new legislation is being introduced in Europe as a consequence of two new Directives specifically aimed at "potentially explosive atmospheres", hence the title "ATEX" from the French "Atmospères Explosives". Directive 94/9/EC on equipment and protective systems[2] and Directive 1999/92/EC on the protection of workers[3] are both concerned with potentially explosive atmospheres ("explosive" is the term used in these official documents for what is commonly called "flammable" in English), but from different perspectives.

7.1 What is in a Name?

There is a lot of confusion about the common names of these Directives, and different names are used. It is common practice in EU circles to designate Directives by the number of the article of the Treaty establishing the European Community that is referred to in the opening statements. So, when there were plans to prepare two Directives relating to potentially explosive atmospheres, the one that was finally published as Directive 94/9/EC was often called "ATEX 100a" and the other "ATEX 118a".

Unfortunately, these designations, although very convenient and used widely in circles such as the standardisation working groups, were never made official. The EU web site refers to Directive 94/9/EC just as "ATEX", but EN 1127-1[4] actually mentions " ... The Council Directive (94/9/EC) ... (called ATEX-100a-Directive) ...".

The confusion was increased when, by the time Directive 1999/92/EC was published, the Treaty of Amsterdam had renumbered the various articles! So, "ATEX 118a" actually in the final version refers to article 137. This means that many people still refer to ATEX 118a, even if that article is not mentioned in the actual Directive, while others have adjusted their terminology and now call it "ATEX 137". Others have gone as far as now referring to ATEX 100a as ATEX 95, after the new number for the "free trade" article, even though that Directive does not actually refer to article 95.

The best solution is probably to refer to "ATEX equipment" and "ATEX worker protection" Directives, or similar designations, but this is almost as cumbersome as using the official designations.

7.2 ATEX 100a Directive

This Directive has been in force since 1996 and has been implemented into national legislation throughout the European Union. Until mid-2003 its application is optional, but after that date it is compulsory.

This Directive applies to equipment and protective systems intended for use in potentially explosive atmospheres, although some safety devices, controlling devices and regulating devices located outside the potentially explosive atmosphere can be covered as well. Essentially, it is a "free trade" Directive, and any equipment conforming to ATEX 100a must be allowed on the market in the EU.

This is a situation similar to the long-established system for "explosion proof" electrical equipment suitable for areas where a gas explosion hazard might arise. In fact, the new system replaces this old one, and very importantly, now covers *all* equipment, electrical and non- electrical, and gas (or vapour) explosions as well as dust explosions and mist explosions.

This Directive is a "New Approach" Directive, which means that only essential health and safety requirements (EHSR's) are laid down. Technical details of how compliance with these EHSR's can be achieved is laid down in European Standards, but it is not compulsory to follow these standards as long as compliance can be achieved.

In the "old" system for electrical equipment, the methods of protection were standardised in European construction standards, but the application of the equipment was subject to national selection and installation rules. This is now changing. The ATEX 100a Directive defines the levels of safety. For equipment of Group II (Group I is for gassy mines), three categories of equipment exist:

- Category 1 has a very high level of protection;
- Category 2 has a high level of protection;
- Category 3 has a normal level of protection (taking into account that the equipment is intended to be used in a potentially explosive atmosphere).

These categories obviously define where the equipment can be used, although that is not stated in the Directive that is only concerned with "placing on the market".

7.3 ATEX 137 Directive

The ATEX 100a Directive has large implication for the manufacturer and for the buyer of equipment, but the ATEX 137 Directive will have a far greater impact on the actual operation of a plant.

Under existing Health and Safety legislation it is already required to operate a plant safely. The new Directive now explicitly demands that the potential explosion hazards are tackled at the source whenever possible. The Directive states that the employer shall take organisational and/or technical measures, in order of priority:

- To prevent the formation of explosive atmospheres, or where the nature of the activity does not allow that,
- To avoid the ignition of explosive atmospheres, and
- Mitigate the detrimental effects of an explosion so as to ensure the health and safety of workers

Where necessary, these measures shall be combined and/or supplemented with measures to prevent the propagation of explosions.

This Directive adds some new elements to the existing practices and the following stand out:

- Organisational measures (e.g. Permit to Work systems, training, maintenance plans and procedures) are as essential to safety as the technical measures (e.g. explosion venting, earthing and bonding measures).
- The first priority should always be the avoidance of the explosive atmosphere, before measures to prevent ignition or to protect plant are even considered.
- Explosion isolation is now very clearly identified as an issue and measures to prevent propagation must be included in the plant.

In line with much other recent legislation, the decisions about plant safety are taken based on a risk assessment. Specifically, the Directive requires that an explosion hazard and risk assessment is carried out, taking into account the special circumstances in the plant under consideration, which includes at least the following factors:

- The likelihood that explosive atmospheres will occur and their persistence;
- The likelihood that ignition sources, including electrostatic discharges, will be present and become active and effective;
- The installations, substances used, processes and their possible interactions;
- The scale of the anticipated effects.

Figure 11 shows the steps that need to be taken in a risk assessment. The same basic steps apply whether an assessment is made of a machine or a piece of equipment (as under the ATEX 100a Directive), or of a process. In all cases it is imperative that a good understanding of the foreseeable process conditions is obtained, including the material properties under those conditions.

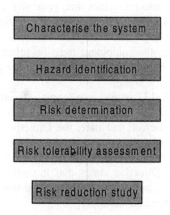

Figure 11 *Steps in risk assessment of equipment and processes*

The first factor above is very closely linked to Hazardous Area Classification, and under the Directive it will become an explicit requirement to carry out a Hazardous Area Classification. In addition, the hazardous areas must be identified in plant by displaying a prescribed sign (Figure 12).

Figure 12 *Sign prescribed by Directive 1999/92/EC to indicate hazardous areas in plant*

With the introduction of the Directive, the hazardous areas for dust explosion hazards will be defined differently. Instead of the two-zone system in use until now (designated Zone Y and Zone Z in most countries), a three-zone system is introduced in parallel with the Zone 0 – 1 – 2 system that is used for flammable gases and vapours. The new definitions are:

- Zone 20 – a place where an explosive atmosphere in the form of a cloud of combustible dust is present continuously, or for long periods, or frequently;
- Zone 21 – a place where an explosive dust atmosphere is likely to occur occasionally in normal operation;
- Zone 22 – a place where an explosive dust atmosphere is not likely to occur in normal operation but, if it does occur, will persist for a short period only.

The results of the explosion hazard and risk assessment, as well as several other aspects, need to be documented in a so-called "Explosion Protection Document". This document must be prepared before commencement of work and must be revised as necessary.

The ATEX 137 Directive also prescribes that, generally, equipment must conform to the ATEX 100a Directive, and states that in Zone 20 only equipment of Group II, category 1 can be used, in Zone 21 category 1 and 2, but that in Zone 22 equipment of category 1, 2 and 3 can be used. Obviously, the equipment must be suitable for the fuel that can be present, which is another reason why the actual flammability properties of the dust must be known.

The time scale for ATEX 137 is very challenging. Even though the Directive was only published early in 2000, the start date will be 1 July 2003 for new plant or subsequently modified plant. Before that date, the Directive must be incorporated into national legislation by all Member States. For existing plant, there is a three-year period before they must comply with the minimum requirements of the Directive.

For work equipment the Directive states that it shall comply with the minimum requirements from 1 July 2003, but if it is already in use before 30 June 2003, then the above mentioned selection criteria do not apply.

7.4 European Standards

It was mentioned before that the technical guidance for compliance with the ATEX 100a Directive will be made available in harmonised European Standards. To achieve this, the European Commission has given a mandate to the European Committee for Standardization (CEN) and the European Committee for Electrotechnical Standardization (CENELEC) to prepare the necessary standards.

In CENELEC, the Technical Committee TC31 has extended the scope of its work to include not only the revision of the existing standards for "Ex" electrical equipment, but also to prepare standards on installation, maintenance and on electrical equipment for use in dust explosion hazardous areas.

In CEN there was no central committee to oversee all work, and TC305 was especially founded for that purpose. TC305 not only considers the construction standards for non-electrical equipment, but also protective systems and, to be able to define the hazards and the protective measures, test methods to determine the explosion characteristics of gases, vapours and dusts. The scope of work in the working groups in

TC305 is extremely wide, and the number of experts available to prepare these standards is very small. Progress is therefore much slower than needed and anticipated. Consequently, many standards will not be available at the time when the users need them.

Although the standards are prepared under a mandate referring to the ATEX 100a Directive, many are also applicable under ATEX 137. For example, to assess the hazards in plant, the material properties are needed, and test standards for many common parameters (explosion severity, minimum ignition energy and temperature, limiting oxygen concentration, etc.) are being prepared in CEN TC305. To assess the adequacy of explosion protection measures, the relevant standards are again needed, for example standards on the sizing and positioning of vent panels.

8 CONCLUDING REMARKS

The analysis of dust explosion hazards in a plant needs to take into account the actual plant conditions and all relevant environmental factors. Seemingly similar plants can thus have a totally different level of hazard and consequently the measures needed to minimise the risk can be different. Likewise, the selection of the most appropriate protective measure can give different results. As there is no preventive or protective measure that is ideal in all respects (product quality, investment, operating cost, environmental impact, etc.), it is important not to exclude any option too soon. In order to select the appropriate Basis of Safety for a plant, all relevant factors must be taken into account, including human factors, the procedures that are in place and possible deviations from the normal operating conditions (see Figure 13).

In general, if the potential dust explosion hazard is recognised before a plant is built, it is possible to improve the design to obtain a better level of safety without additional

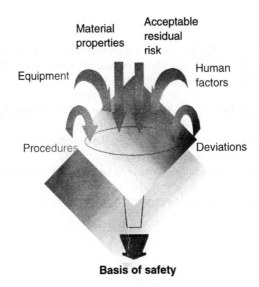

Figure 13 *Factors influencing the selection of an appropriate Basis of Safety*

investment. Besides, any investment that is still needed will be lower than if explosion prevention and protection are only considered once the plant is built.

The new ATEX Directives provide a structure to assess the hazards and risks and to design preventive and protective measures to obtain a level of risk that is as low as is reasonably practicable. The underlying methodology of the ATEX Directives is not new, but their implementation will certainly bring about some changes in industry. Early recognition, assessment and reduction of explosion hazards has not been implemented in the past consistently in some parts of industry. These industry sectors should see benefits in terms of safety as return for the additional effort that is undoubtedly needed to comply with the ATEX Directives.

References

1. A. Fowler and J. Hazeldean, *Fire Prevention*, 1998, **311,** 22.
2. *Directive 94/9/EC of the European Parliament and the Council of 23 March 1994 on the approximation of the laws of the Member States concerning equipment and protective systems intended for use in potentially explosive atmospheres*, Official Journal of the European Communities, L100, Volume 37, 19 April 1994.
3. *Directive 1999/92/EC of the European Parliament and of the Council of 16 December 1999 on the minimum requirements for improving the safety and health protection of workers potentially at risk from explosive atmospheres (15th individual Directive within the meaning of Article 16(1) of Directive 89/391/EEC on the introduction of measures to encourage improvements in the safety and health of workers at work)*, Official Journal L23, 28 January 2000.
4. EN 1127-1:1997, *Explosive atmospheres – Explosion prevention and protection – Part 1: Basic concepts and methodology,* CEN, March 1997

Bibliography

There are several useful handbooks on dust explosions. Some of these are listed below. More detailed information on various aspects of dust explosions and explosion prevention and protection are published in specialist journals.

W. Bartknecht, *Dust Explosions - Course, Prevention and Protection,* Springer Verlag, 1989.
W. Bartknecht, *Explosionsschutz - Grundlagen und Anwendung*, Springer Verlag, 1994.
J. Cross and D. Farrer, *Dust Explosions,* Plenum Press, 1983.
R. K. Eckhoff, *Dust Explosions in the Process Industries*, Butterworth-Heinemann, 2nd edition, 1997.
P. Field, *Dust Explosions,* Elsevier Scientific Publishing Co., 1982.

PRACTICAL SOLUTIONS TO CRITICAL SOLIDS HANDLING PROBLEMS

P. Cooper

Matcon Limited
Matcon House, London Road
Moreton In Marsh, Gloucestershire GL56 0HJ

1 INTRODUCTION

Matcon is a world leading supplier of batch orientated powder handling systems based on the Intermediate Bulk Container (IBC) concept. This paper analyses the history of these concepts and some of the possible areas of use within many process industries. We will also describe the Principal of Operation of the Matcon IBC System and highlight different flow patterns when discharging from IBC's and pro's and con's thereof. We also describe a number of application examples and their benefits.

2 HISTORY OF BATCH POWDER PROCESSING

In the '60s process industries such as food and pharmaceuticals, began using IBCs which were simple, often home produced boxes to transfer bulk materials between different processes (Figure 1). This approach allowed flexibility of production, but had severe limitations because the filling and discharging systems available were dusty, could not discharge some materials and could not control the flow of product to the process below.

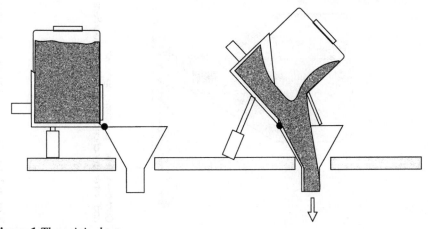

Figure 1 *The original tote*

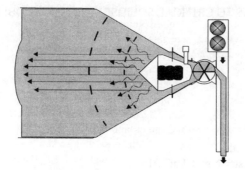

Controlled feed to rotary valve and pneumatic conveying system.

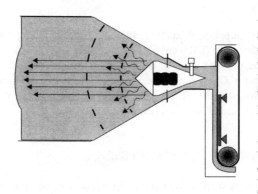

Controlled feed to weigh feeder/belt feeder

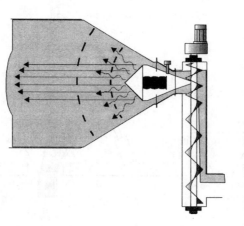

Controlled feed to screw conveyor and other mechanical conveying equipment.

Figure 2 *Fixed hopper and pneumatic conveying systems*

These limitations together with a trend in the '70s and '80s towards larger and longer production runs, ("big is beautiful"), caused the industry to move towards fixed hopper and pneumatic conveying systems as a means of transferring bulk materials (Figure 2).

The advantages of this approach were that no make and break connections were necessary during processing and it offered a degree of automation. However suitable this concept in continuously operating chemical plants, it has limitations in batch processing systems. It limits the flexibility of production; cross contamination is inevitable; segregation of mixed products is likely during conveying; and frequent cleaning is required, leading to significant downtime and escape of dust.

In the '90s the trend turned again both regarding production equipment and philosophy. The "buzz words" of today are "just in time" and "custom fit products" justifying higher prices and improving profitability. The IBC Cone Valve Technology invented and perfected by Matcon, overcomes the problems experienced using earlier IBC systems as well as overcoming the limitations of continuous systems (Figure 3). Not only has IBC technology improved dramatically, but the use of bar codes and improved transportation systems now allows batch traceability and automation which previously wasn't feasible.

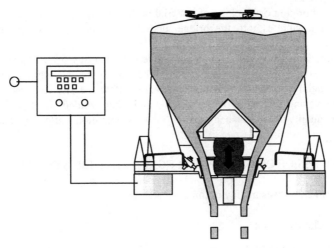

Figure 3 *Cone valve IBC system*

3 DISCHARGE FROM IBCS/SILOS, CORE (FUNNEL) FLOW – MASS FLOW

Any hopper, silo or container achieves a "pattern" of flow when discharging solids.

These flow patterns can sometimes be influenced by design of the storage vessel or by a discharge device which controls the flow from the vessel.

Most flow patterns from storage vessels fall into the category of <u>Core (or Funnel) Flow</u>. This is when flow occurs in a narrow vertical flow channel from the outlet to the top surface of the material. The material on the top surface will always prefer to flow towards the centre with the central downward flow applying pressure to hold the material being stored along the sides of the vessel.

Although there is nothing wrong with vessels storing unblended, individual solids (flour, grain, etc) discharging on a Core (or Funnel) Flow basis, there is an inherent problem with this type of flow when discharging blended solids. This is because the flow pattern induces separation of fine and coarse particles as they free fall through the core at different rates, as well as a tendency for the fine particles to discharge first. The discharge of any blended product from a vessel or blender where segregation must be avoided, should achieve a <u>Mass Flow</u> pattern. This is when the material discharges evenly across the vessel without any preferential flow taking place. This ensures that the material "lowers" in the vessel at a consistent rate avoiding any rush or free fall of solids from occurring.

A mass flow design of vessel will usually use steep angles in the hopper section of the vessel so that the material will prefer to flow down the hopper walls than to core through the centre of the vessel. Cone Valve IBC Technology prevents any form of core or funnel flow from occurring by preventing any flow through the centre of the IBC and directing flow through an annular gap around the Cone Valve and towards the sides of the IBC. Flow can be induced and controlled by raising/lowering and vibrating the Cone Valve during discharge.

The risk of bridging or blocking of material in the IBC is also overcome as the Cone Valve raises and vibrates into the bridge, undermining its compressive strength and causing it to collapse.

Having solved the potential segregation problem discharging from a silo or an IBC, it is imperative to avoid the problem occurring on the way to the packing machine. Ideally the packing machine should be positioned directly below the silo/IBC and high speed pneumatic conveying (vacuum or pressure) should be avoided as it will inevitably deteriorate the quality of the finished product (i.e. by segregation and/or degradation).

4 PRINCIPLES OF OPERATION FOR CONE VALVE IBC SYSTEMS

The IBC System comprises mainly of three parts:-

- The IBC itself
- The Filling System
- The Discharge System

The three parts and the principle of operation are described below.

4.1 Intermediate Bulk Container

The typical Intermediate Bulk Container (IBC) is square in cross section with a square to round transition hopper ending in a circular outlet, in which is located an inverted cone with an integral edge seal. This becomes the "Cone" Valve and gives a dust tight seal under all conditions of full or empty IBC and during any form of transportation of the IBC.

Typical IBC sizes range from 500 litres up to 2,500 litres. They are manufactured completely in stainless steel and plastic or plastic bottle inserted into a robust steel cage (Figure 4).

Figure 4 *Typical Intermediate Bulk Containers*

4.2 The IBC Discharge Station

The typical IBC Discharge Station consists of a square frame with corner guides to ensure that the IBC is automatically positioned centrally on the Discharge Station (Figure 5). There is also a central hopper with an integrated lifting/lowering probe.

The Cone Valve remains closed during loading and the IBC hopper section engages a lip seal on the discharge hopper to give an automatic dust tight seal between the two units each and every time.

There are a large variety of options available for this basic unit, including weighing systems, internal and external vibration, extra lift force and height and extreme containment using air washing.

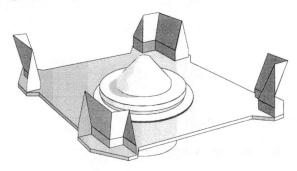

Figure 5 *Typical IBC discharge station*

The basic principle of operation is however always as described below giving the following main advantages:-

- A controlled feed to the following process preventing compaction
- Mass flow discharge preventing segregation
- No risk of flushing as the Cone Valve always stops the flow when in the seated position
- Dust free operation both during and after discharge

- Possibility to perform dosing direct from the IBC
- Capability to handle virtually any powder materials

4.3 Principle of operation

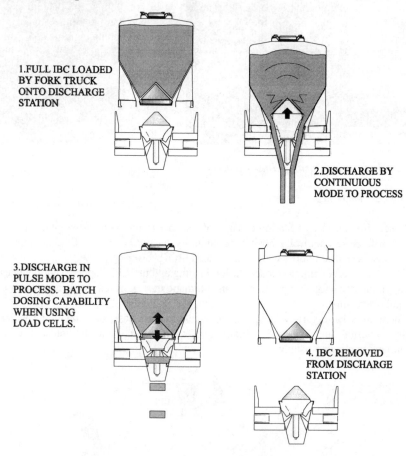

1.FULL IBC LOADED BY FORK TRUCK ONTO DISCHARGE STATION

2.DISCHARGE BY CONTINUIOUS MODE TO PROCESS

3.DISCHARGE IN PULSE MODE TO PROCESS. BATCH DOSING CAPABILITY WHEN USING LOAD CELLS.

4. IBC REMOVED FROM DISCHARGE STATION

Figure 6 *Principle of operation of IBC*

4.4 Containment Transfer System (fully automatic)

The Containment Transfer System (CTS) enables transfer of powders from a process to an IBC while maintaining containment of the material from the outside environment. There is not even a momentary exposure of the inside of the IBC to the outside before, during or after the transfer. The system is totally automatic with no manual intervention whatsoever. The operator does not have to remove or replace a lid or operate a valve.

The system consists of:-

- A containment unit to bolt or flexibly connect to the bulk store or process vessel outlets
- A loose lid inserted in a lip seal arrangement in each IBC

- A separate control cabinet

The principle of operation is clearly described in Figure 7, the engagement between the CTS and the IBC can be performed either by lifting the IBC or indeed lowering the CTS which Matcon supply with a cylinder mechanism as a standard option. When following the sequence of operation it is particularly important to notice how the IBC lid and the underside of the Cone Valve in the CTS are engaged for as long as transfer takes place and are only separated once the transfer process is finished. Thereby extreme containment levels of down to less than 10 micro grams per cubic metre can be achieved, even without secondary dust extraction.

5 MIXING

One of the most significant functions and consequently one of the most important aspects in a process system is blending. In most food applications it is the major source of increasing value to the products that go into the process. The quality of the mix performed within the blender as well as when transferred to the next process, typically packing, is the single most important consideration for project managers and quality assurance personnel.

5.1 Blending Process methods

Regardless of blender type there are two primary process characteristics recognised in powder blending theory, namely **'Flow'** and **'Shear'**. Both of these are present, in different proportions, in every blender type. It is the application and control of the appropriate combination of Flow and Shear that leads to efficient blending.

Flow refers to the relative movement of particles within a material. The ability to randomly create that movement of particles encourages intermeshing and blending. High flow and a good blend will be achieved with materials that do not stick together. Flow is maximised and typically induced via tumbling or to a lesser extent via an internal conveying type device such as spiral blades, paddles and ploughs/plows (Figure 8).

Shear is the amount of work applied to the material forcing the product particles to move. Applying the correct level of shear to particle forces intermeshing and blending. Shear is a necessary component when blending materials that tend to stick together and agglomerate, and is especially important when blending with liquids. However if mismanaged, particle degradation and heat build-up may occur. Shear is typically induced via a high speed, high impact internal blade mechanism, such as chopper blades or impellers (Figure 8).

5.1.1 After mix considerations. Ensuring efficient blending requires many qualities in a blender, not only the ability to blend effectively, thoroughly and quickly but also aspects such as the cleanability, dust tight operation, ease of filling and discharging and handling of the material.

Each aspect is as important, for although a blender may blend thoroughly and efficiently, if the method of discharge is prone to segregation or the blend is handled frequently, also causing segregation, the blend quality is affected. Discharging the product correctly with minimum handling is therefore as important as blending efficiency.

In small to medium sized operations it has been common practice to perform the packing into 25 kg bags direct from the static mixer outlet. This is very inefficient as it

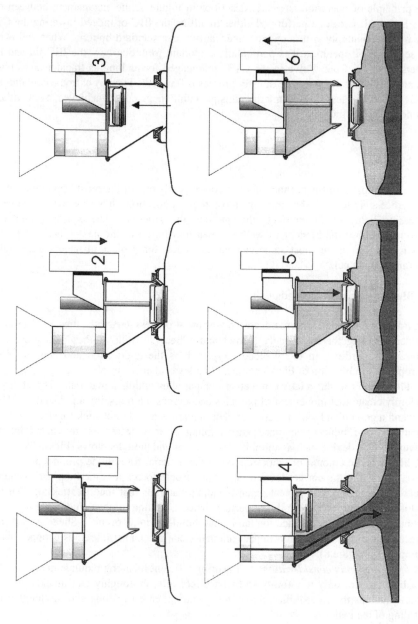

Figure 7 *Principle of operation of containment transfer system*

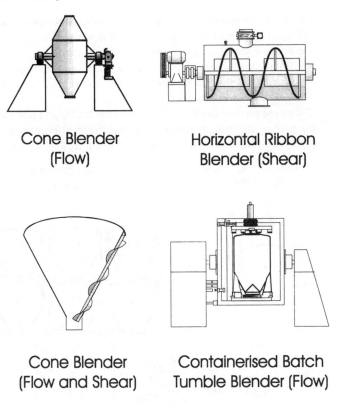

Cone Blender (Flow)

Horizontal Ribbon Blender (Shear)

Cone Blender (Flow and Shear)

Containerised Batch Tumble Blender (Flow)

Figure 8 *Examples of blending process methods*

occupies the blender during the complete packing and there is also risk of overworking the material causing degradation. Therefore there is a requirement to discharge the mixed batch from the blender down to a storage hopper or an IBC. During the dropping of the mix batch from the blender to the storage hopper, there is a definite risk of segregation of the blend. This is especially true in the case of dry lose particulate solids requiring high flow, low shear type mixing.

5.1.2 Containerised Batch Tumble Blender. The Containerised Batch Blender does not have the problem mentioned above because the mix batch is not dropped. Instead it remains in the Intermediate Bulk Container (IBC) while it is removed from the blending cage, transported direct to the packing line and discharged or stored until required (Figure 9). Other significant advantages of the IBC blender are:-

- Once an IBC has been removed from the mixing cage, the blender is free to mix the next batch without cleaning.
- The removable IBC eases/simplifies the quality assurance, by avoiding cross contamination as separate IBCs are ensuring batch integrity. The mixed product can be retained in a IBC during quality assurance always ensuring a repeatable process without the cost of going through the packing process.

- The blender does not need to be cleaned between different batches. Each IBC can be taken away from the mixing/process site and be washed down allowing continued production in the blender.
- No filling or discharge of the mixer is required as this is performed elsewhere into the IBCs. Besides the actual time saving it gives a complete dust free operation.
- The mixed product can be moved within the IBC to an ideal position above the packing machine avoiding the pneumatic transfer that may cause both segregation and degradation of the product during the transfer.

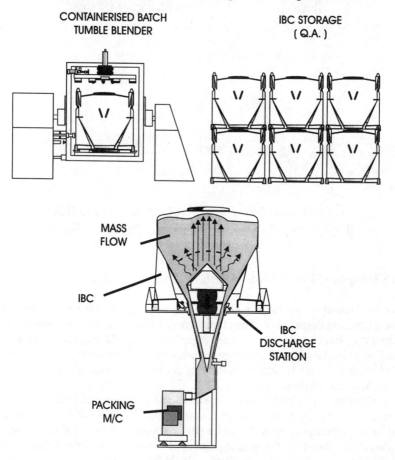

Figure 9 *Containerised batch tumble blender*

5.2 Principle of Operation

IBC blenders are available from laboratory scale up to 1,500 litre gross volume, even as multi-size if required. Both steel and plastic IBCs can be mixed. The IBC is simply inserted on an angle into a cage and then vertically rotated on an asymmetric axis creating an irregular flow pattern, moving all particles within the IBC. The angle has been defined through excessive testing and should not be changed whilst the rotation speed varies depending on the size of the IBC and product to be mixed. To achieve the highest possible mixing throughput, the IBC should be filled to approximately two thirds full.

5.2.1 Interprocess Matcon IBC System. The use of the Matcon IBC System to transfer finished blends from the batch mixer to the packing machine is now regularly employed in the industry. The detachable IBC makes the mixer itself a lot more flexible, the batches can be campaign mixed, unloaded from the mixer and stored in the mixing IBC, retaining its mix until required. The mixer is then ready for mixing the next batch or to be cleaned.

The potential risk of dust and cross contamination when making and breaking connection between a mixer and the IBC inlet can be overcome using the Matcon CTS. Kerry Ingredients applied this concept from 12 mixers all charging 2.5 cubic metre Matcon IBCs. Containment levels of less than 2mg per cubic metre have been achieved using this principle (Figure 10).

Filling and discharging is often underrated in its importance when thinking of a mixer. Yet as briefly mentioned earlier, discharging especially, can detrimentally affect

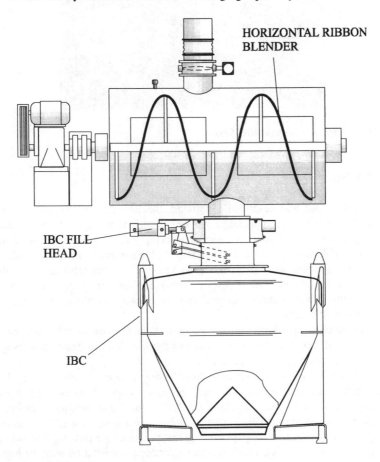

Figure 10 *Interprocess Matcon IBC System*

the quality of mix. With a detachable mixing IBC, filling can be carried out in a separate area at floor level, the mixing IBC can then be taken and easily loaded into the mixer, reducing any dust contamination. Equally so, discharging can be carried out in a separate area directly above the packaging machine hopper (Figure 11). The IBC with material can either be stored until required or discharged direct to packaging. Either way, the product is not handled excessively and the mix is thus not prone to segregation or degradation – the mix quality is maintained.

Figure 11 *Interprocess Matcon IBC System*

6 PACKING

As already mentioned and described in mixing, the Matcon IBC System is the ideal transfer system from mixing to packing. Not only that, it also helps the packing machine by providing a controlled flow. Its normal working mode is to employ level sensors controlling the refill making the Cone Valve raise and lower as and when required.

An interesting aspect when using the Cone Valve IBC as the refill device, is that the horizontal screw feeder can in most cases be removed from the system. The refill from the IBC is controllable to such a degree that the packing machines accuracy can be withheld without this screw.

The same applies for filling systems using horizontal screw or vibratory feeders as dosing devices. Matcon have integrated many of these systems to complete packages for small throughput packing (Figures 12 & 13).

In multi-purpose plants where both free-flowing and extremely difficult flowing cohesive powders are being handled, the volumetric refill operation described above may not be adequate. In these instances the same PLC controller as mentioned earlier under formulation can be employed performing a continuous dosing as a refill. Instead of requesting a batch, a certain throughput rate to match that of the packing line is keyed in by the operator. The two systems will then operate in conjunction with the high level sensor as a secondary security to shut off the Matcon in emergencies. Kerry Ingredients, for example, handles approx. 2,000 different mixes and have commissioned the Matcon

Figure 12 *Filling system using horizontal screw feeder*

Figure 13 *Filling system using vibratory feeder*

Loss In Weight System for each of these products, which it handles extremely successfully.

NEW DEVELOPMENTS IN THE CONTAINMENT OF PHARMACEUTICAL POWDERS

Martyn Ryder
Extract Technology Ltd
Bradley Junction Industrial Estate
Leeds Road
Huddersfield HD2 1UR

1 INTRODUCTION

There is a growing trend for routine day to day pharmaceutical powder processing and production operations to involve "high potency" powders. To meet this challenge new strategies for powder handling are being developed along with engineering advances in the equipment used to contain the active or high potency materials. The term containment of course refers to the isolation of hazardous materials (high potency) using engineering methods in various guises. Containment can protect the operator from the compound or protect the compound from operator born contamination. In some cases specialised containment devices protect the operator from the materials and vice-versa.

First of all let us look at the growth of high potency compounds within the pharmaceutical industry. In 1990 only about 5% of all pharmaceutical actives handled were considered potent. Now in the Year 2000 up to 30% of all API's are classified as potent and therefore require special attention to protect the operator from the effects of coming into contact with these materials.

Table 1 provides a chart from American Industrial Hygiene Association Journal which classifies the enrolment criteria for potent compounds. Hazard Group A, not shown on this chart, are excipient materials. Hazard groups B and C are semi-potent compounds, but hazard groups D, E and F are seriously considered to be potent compounds and necessitate extreme caution and care when these materials are introduced into the pharmaceutical facility.

In addition to considering the health of process operators the pharmaceutical manufacturer also needs to reflect that the true goals of current Good Manufacturing Practice (cGMP) when applied to a pharmaceutical manufacturing facility include fitness for purpose in that it must comply with regulatory standards and be consistent in the quality of material produced.

Without effective containment programmes in place cross contamination within the manufacturing facility is a serious possibility which cannot be overlooked. Hence from a quality assurance perspective effective containment of potent compounds is one of the key factors to be considered when applying the goals of cGMP to our facility design philosophy.

Table 1 *What is a Potent Compound?*

Quoted from the American Industrial Hygiene Association Journal

Equipment Criteria	Hazard Group B	Hazard Group C	Hazard Group D	Hazard Group E	Hazard Group F
Potency (µg/day)	>100	10-100	1.0-10	1.0-0.1	<0.01
Severity of acute (life-threatening) effects	low	low/mod	moderate	mod/high	high
Acute warning symptoms Onset of warning	good immediate	fair immediate	fair/poor may be delayed	poor delayed	none none
Symptoms medically treatable	yes	yes	yes	yes	yes/no
Need for medical intervention	not required	not required	may be required	may be required immediately	required immediately
Acute toxicity Sensitisation	slightly toxic not a sensitiser	moderately toxic mild sensitiser	highly toxic moderate sensitiser	extremely toxic strong sensitiser	super toxic extreme sensitiser
Likelihood of chronic effects (eg: cancer, repro, systemic)	unlikely	unlikely	possible	probable	known
Severity of chronic (life-shortening) effects	none	none	slight	moderate	severe
Cumulative effects	none	none	low	moderate	high
Reversibility	reversible	reversible	may not be reversible	may not be reversible	irreversible
Alteration of quality of life (disability)	no	no	yes/no	yes	yes

2 PERFORMANCE EVALUATION – CONTAINMENT DEVICES

As is well known the pharmaceutical industry is highly ethical and will not market a drug compound unless its efficacy, side effects and overall systemic operation are fully understood. In line with our customers philosophy, in 1989 the Board of Directors of Extract Technology decreed that Extract Technology must *"test - challenge and evaluate the performance of its containment products publicising test data."*

For this reason we have continued a programme of intensive performance testing since 1989. These have covered cross contamination assessments when operating with a recirculatory downflow booth, looking at operator exposure in great detail when dispensing powders, and considering the microbiological burden that a containment booth or isolator might carry after many years of operation. The test programme has also considered the effectiveness of barrier isolation and Clean In Place (CIP) technology used to decontaminate isolator chambers.

Finally in the Year 2000 we have completed an evaluation of split butterfly valve containment and surface contamination together with a correlation study. This study evaluates the accuracy of real time dust in air monitoring techniques which are necessary for engineers designing containment systems with the industrial hygiene favoured methods of hunt sampling within the operator's breathing zone.

All of this test data has provided us with a very broad understanding of the typical performance criteria that a number of containment devices may provide. This led to the development of our very first 'Pyramid Chart' in 1992 that categorised different types of flow booth and isolator technology according to their approximate level of operator protection (Figure 1).

3 CONTAINMENT STRATEGY BASED SELECTION

Since 1996 Extract Technology has been involved in the production of a Design Guide for containment technology in co-operation with the British Institution of Chemical Engineers. During this work we have also been involved with the British Health and Safety Executive. The benefit of this interaction with these key organisations has accelerated our understanding and now drives us towards setting "containment strategies" as the method we will use for approaching a containment project with one of our clients (Figure 2).

In our 'containment strategy' we think about the big picture. The evaluation team study the process in detail particularly the operator's interface with the potent compounds to be handled. These may be solids or liquids but in any event we want to understand how much interaction the operator will have with the materials and how dusty or volatile these materials will be. We also need to look at the hazards created by these materials and their exposure limit or Operator Exposure Limit (OEL). Table 2 provides a hazard group banding system from A to F that will be used in the forthcoming I.Chem.E. Guide.

The task specific exposure risk and hazard combine to drive us into one of the key containment strategies recognised by the Health and Safety Executive (HSE) and I.Chem.E. Before we look at what the containment strategies generally cover we must understand that a containment strategy is not merely the selection of a suitable device of a given type. A containment strategy considers the essential operator training, development of standard operating procedures, access training and emergency training together with maintenance and cleaning protocols that are needed to ensure that the containment device itself continues to operate satisfactorily and within its design parameters.

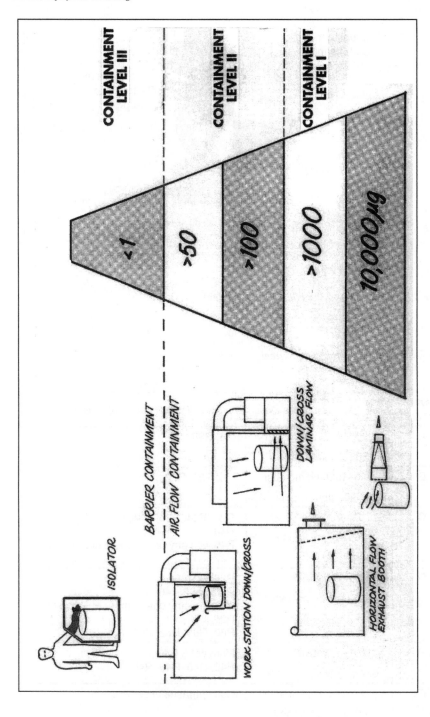

Figure 1 *1992 Pyramid chart*

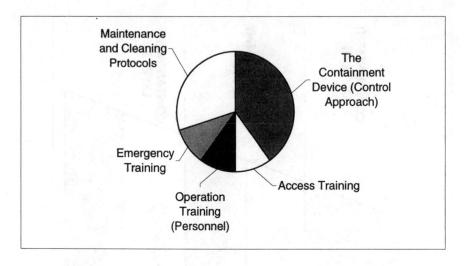

Figure 2 *Key components of a containment strategy*

Table 2 *Hazard Group Banding System*

Hazard Group	Operator Exposure Limit	Typical basic R-phrases
A	1000–10 000 µg m^{-3} dust 50–500 ppm vapour	R36, R38
B	100–1000 µg m^{-3} dust 5–50 ppm vapour	R20, R21, R22
C	10–100 µg m^{-3} dust 0.5–5 ppm vapour	R23, R24, R25, R34, R35, R37, R41, R43 R48 with any one or more of R20, R21, R22
D	1.0–10 µg m^{-3} dust 0.05–0.5 ppm vapour	R26, R27, R28 Carc cat 3 R40 R48 with any one or more of R23, R24, R25 R60, R61, R62, R63
E	0.01–1.0 µg m^{-3} dust 0.005–0.05 ppm vapour	Muta cat 3 R40 R42, R45, R46, R49
F	<0.01 µg m^{-3} dust; <0.005 ppm vapour	No R-phrases assigned.

Five key strategies have been identified by the British Institution of Chemical Engineers.

Containment Strategy 1	Controlled General Ventilation
Containment Strategy 2	LEV (local exhaust ventilation) and airflow booths
Containment Strategy 3	Barrier isolation systems whereby **open powder transfer** within the isolator chamber itself is permitted
Containment Strategy 4	Barrier isolation systems with contained powder transfer systems or packages within them
Containment Strategy 5	Fully automated or robot type operation
	No direct operator involvement

The differences between Containment Strategies 3 and 4 have been brought about by the realisation that the weak links in any barrier isolation system are the inlet and outlet transfer doors/valves/ports. These could be Double Porte de Transfert Entaché (DPTE®) ports, split butterfly valves or similar. When these ports are contaminated with the drug compound it is inevitable that some break out of material will occur due to either poor maintenance or ingress of the powdered material itself into the finely machined faces of these transfer ports. As such this brings about an unacceptable risk of powder break out. Containment Strategy 4 eliminates this risk by requiring that all powder transfers within the isolator body are fully contained. Here the isolator provides an outer layer of containment and an extra level of safety for our operators.

So how do we use the Containment Strategy principle to select the right approach to a containment project? The first step is to evaluate the operator's exposure to dust or vapour for the client's specific operation.

Exposure Potential (EP) is the key factor that drives the risk associated with a specific operation. In our new pyramid chart we have a table which quantifies the quantities of material being handled in the specific process and their dustiness or volatility. These figures drive us to an exposure potential ranging from EP1, the lowest level of exposure potential, to EP4, the highest level of exposure potential.

Exposure potential is dependent (for solids) on two key factors; the dustiness of the material and the amount of material being handled. Tables 3-5 show the relationship between these factors and exposure potential.

As may be seen from the worked examples that follow, the exposure potential drives the risk assessment part of this initial stage of our project evaluation. Once we have the exposure potential the pyramid chart (Figure 3) permits us, simplistically, to cross reference the hazard band of the materials being used with the exposure potential and leads us to a recommended containment strategy. Overall the pyramid permits us to make a simple **yes** or **no** conclusion as the suitability of a typical containment device within a few minutes of finding the facts and figures associated with the project.

The following worked examples prove the above point.

In worked example 1 we consider sampling pharmaceutical active materials in a supplies inspection area of a warehouse. Obviously small quantities of material are being handled and very careful powder transfer is being used to put small samples into a container for QA analysis, hence the exposure potential is EP1. From the material data sheet (MSDS) the OEL is 15 microgrammes, this indicates hazard band C. Exposure

Table 3 *Dustiness Potential Assessment*

Dustiness	Characteristics
Low	Pellet like, non-friable solids Little dust seen during use *Examples: tablets, steroids*
Medium	Crystalline, granular solids Dust is seen during use but settles out quickly Dust is seen on surfaces after use *Example: Wet grain*
High	Fine, light powders When used, they form dust clouds that can be seen and that remain airborne for several minutes *Examples: Milled bulk drug*

Table 4 *Scale of Operation Assessment*

Operation	Scale
Lab/pilot transfers – gm–kg (solids) or 1–1000 millilitres (liquids)	Small
Transfers in range 10–100 kg (solids) or 1–1000 litres (liquids)	Medium
Large-scale production transfers – over 100 kg (solids) or above 1000 litres (liquids)	Large

Table 5 *Exposure Potential Based upon Dustiness Potential and Quantity Handled*

Dustiness potential / Quantity handled	Low	Medium	High
Small (gm)	EP1 / EP1	EP1 / EP2	EP2 / EP3
Medium (kg)	EP1 / EP2	EP2 / EP3	EP3 / EP4
Large (tonnes)	EP2 / EP3	EP3 / EP4	EP3 / EP4

Task (transfer) duration:	Short / Long	Over 30 minutes considered long

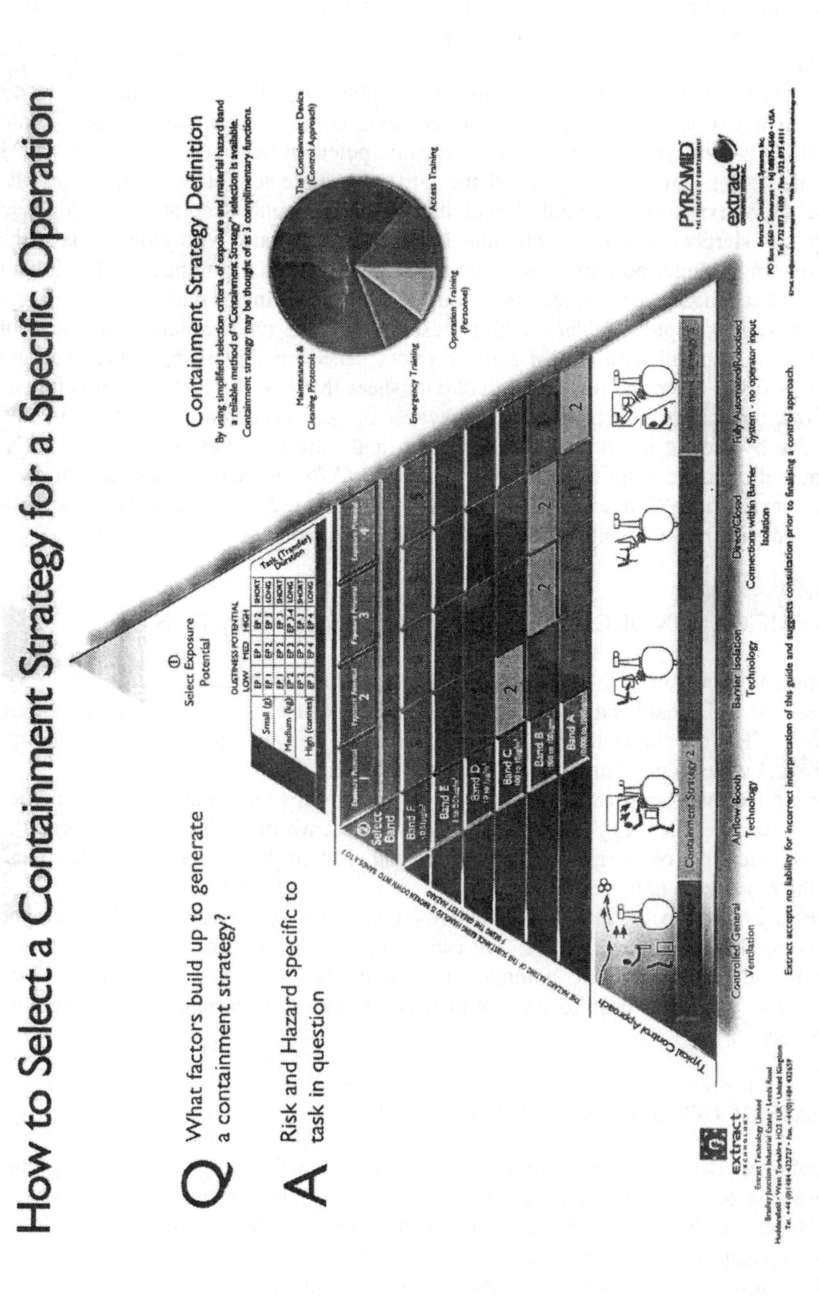

Figure 3 *Pyramid chart*

potential and hazard band C intersect on Containment Strategy 2 which is suitable for a laminar airflow or a clean air booth to provide the operator protection during this operation.

In worked example 2 our particular process involved the filling of 57 kilo drum packs from the outlet of a process dryer with a very high potency compound. The 57 kilo drums of a dusty material put us into the exposure potential band of EP3. The MSDS for this particular compound suggested the OEL was quite low at $5\mu g/m^3$. Using the pyramid chart exposure potential 3 and hazard band D intersect at a Containment Strategy 4. Here we select a continuous liner pack-off isolator with glove ports at the front providing a direct powder connection around the process filling head. The Isolator becomes the secondary containment and should not be contaminated during drum filling.

In worked example 3 tablet friability testing we have much smaller quantities of material being handled with almost zero dustiness hence for this example the exposure potential is EP1. From the material safety data sheet (MSDS) the material OEL for the active compound is $2\mu g/m^3$ or band D. However the active content in the table is only around 2% by weight so the customer's Health and Safety Department allowed us to downgrade the hazard band from band D into band C for the active excipient mixture. EP1 and hazard band C intersect on Containment Strategy 2, so once again an airflow protection device like a downflow laboratory bench is suitable for this operation.

4 VERIFICATION OF CONTAINMENT STRATEGY SELECTION

It is critically important at the point of containment strategy selection to cross reference our selection with health and safety professionals, the user group and the validation department. Table 6 shows the recommended interface between the various user groups to ensure that the containment strategy selection is correct.

Having received approval from the various user groups and other interested parties that the containment strategy selection is indeed correct we may continue to look at the various generic types of barrier isolation system and clean air booth that are marketed and look at the key design points of each design.

One key point to remember is that whilst volumes of powder handled and the dustiness potential are "facts of life" we can minimise the operators interface with the powder by adopting simplistic automation methods eg. gain in weight powder feed systems, drum tippers or post hoists - all of which reduce the operator's interaction with the materials.

5 EASE OF OPERATION - PRODUCT FLOW

All containment devices irrespective of their design or manufacture place a burden on the operator's ease of use. The more we control the operator's position and interface with the materials the less easy the systems are to operate but we have the benefit of reduced operator exposure and therefore reduce the risk to the operator's health.

When designing the powder handling zone special attention should be paid to materials management and personnel flow. Inflowing materials are ideally entered from the warehouse to one side of the dispensing cubicle with weighed assemblies going into batch containers or batch cages at the opposite side. Removal of waste materials particularly in areas of high volume subdivision is a serious issue and must not be overlooked.

Table 6 *Recommended Interface between the Various User Groups*

Elements \ Input	Chemical Supplier	Occupational Hygienist	Health & Safety Development	Operations	Maintenance Group	Quality Assurance Validation	Equipment/System Designer
Hazard Group	√	√	√				
Scale of operation				√			
Exposure potential		√	√	√	√		
Frequency & task duration				√	√		√
Operability of device			√	√	√	√	√
Cost of device		√		√	√		√

As the potency of the materials being handled increases we will be driven inevitably into the area of barrier isolation technology to provide adequate levels of operator protection. Here new challenges await as the ergonomic interface between varying heights of operators often proves very difficult to overcome. One key design recommendation the author proposes is the provision of a full scale timber mock up onto the manufacturing site to permit the user group to carry out functional evaluation of the isolator shape and to ensure that the task actually can be carried out in the manner that the designer has proposed. It is far easier to adjust the mock up of course than the final stainless steel design.

Airflow systems within barrier isolation units vary greatly from vendor to vendor. We have data available that shows that high volume airflows within the isolator actually minimise contamination of the internal surfaces and reduce the cleaning burden at the end of the campaign irrespective of the airflow system used within the isolator. Some method must be included in the design to permit decontamination of the isolator chambers. The techniques chosen may include manual wiping with solvent solutions - manual wash down with a hand held spray gun - or - a fully automated wash down system (Clean In Place, CIP).

Where clean in place technology is required to decontaminate the isolator it is in our experience very difficult to predict accurately the cleaning potential of any spray ball systems at the design stage. Our philosophy is now always without fail to carry out a cleaning evaluation test during the Factory Acceptance Test (FAT). Here the isolator may be contaminated with a suitable marker compound subject to the CIP cycle and the level of cleaning evaluated by swab testing.

Containment of potent compounds in the pharmaceutical industry is now drawing on new technology ideas to permit easier and more cost effective methods of containing materials handled during process operations. One example of this is the development of the split butterfly valve system. The split butterfly valve pioneered by Glatt Systemtechnik, Germany, Serck in the UK and Buck in Germany permits the docking of a pre-charged container onto a process vessel without the need for a second isolator.

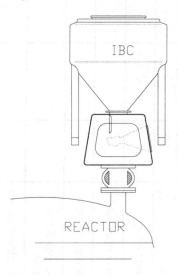

Figure 4 *Rapid Transport Port (RTP) type port connection*

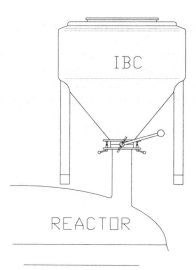

Figure 5 *Split butterfly valve connection*

From Figures 4 & 5 it may be seen that by comparison when using a conventional alpha/beta port (DPTE®) each receiving vessel needs its own small isolator glove box to remove the port doors to permit the flow of powder from the transit container into the open vessel.   The split butterfly valve technology eliminates this secondary glove box and therefore makes the process far more economical.

To provide a cost effective "upgrade" to split butterfly valve technology Extract's R & D team developed the Microcharge containment. This holds the split butterfly valves a distance of 100mm apart after powder transfer. The airflow through the containment prevents contamination break out during valve separation and manual hand wiping of the valve faces with a solvent saturated wipe. This system was the result of many months intensive airflow study using Computational Fluid Dynamics (CFD) to ensure no possibility of reverse flows within the containment during cleaning operations. The goal of this project was to establish if the usual containment limit for a split butterfly valve, ie circa $10\mu g/m^3$ dust in air level, could be bettered by using improved airflow. It is anticipated that the microcharge system will provide a containment level of below the 1 microgramme limit with the correct airflow and operational procedures.

Another growing area of interest in potent compound containment is in the use of "disposable technology" glove bags and smart liners.   ILC Dover in the USA have developed their Doverpack Bag to permit virtually contamination free discharge of process vessels and filter dryers without the need for a rigid glove box.

In our own area of expertise we have adopted glove bag technology to provide low cost localised containment for pilot scale or laboratory operations such as small scale granulators and mixers etc. Just as with barrier isolation system the key to success when employing glove bag technology is to produce and trial a prototype bag with the user group to ensure that its operation will not hinder the process.

6 CONCLUSIONS

There are many popular types of containment technology available in today's

Pharmaceutical Industry. These may range from disposable systems, glove bags or airsuit Personal Protective Equipment (PPE) - to the most sophisticated automated handling systems. Often the simple "clean air booth" can provide good levels of protection without the burden of investment and difficult ergonomics of glove box isolator systems. Selection of the correct system can be made easier by using the Pyramid Chart. This provides a good insight to the containment strategy and type of device needed to get a right first time solution.

New Technologies

POWDER PRODUCTS AND STRUCTURE

John Dodds

Ecole des Mines d'Albi-Carmaux
Campus Jarlard
81013 Albi
France

1 INTRODUCTION

Making solid products implies mixing together powder ingredients and their assembly in some sort of structure such as granules, extrudates, tablets or even free powder mixtures, which will provide the end use properties required by consumers. As in the formulation of liquid or emulsion products, the primary requirement is to establish the correct dose of each ingredient for the required function and to ensure their mutual chemical compatibility. The second stage is to assemble the ingredients in a structure, which will provide the end use properties and efficiency for the required function combining stability, handling, dispersability, etc. This procedure for product design is usually based on empirical methods and strategies for new products usually involve minor modifications to existing formula making heavy reliance on experience. This means that formulation methods are difficult to transmit to others, are non-conducive to innovation and leave little place for acquiring an ordered body of information about products. Another consequence is that formulas tend to become more and more complex involving more and more components, the overall utility of which are difficult to establish.

It may be supposed that, as in other fields of technology, a more analytical approach to solids formulation could be useful. That is by applying the scientific method of simplifying complex systems to find their inherent substructure it may be possible to obtain some sort of framework which would lead to more understanding and permit generalisations in product design. It is probable that practical formulations involving say a dozen or so components may pose too complex a problem for simple analysis, in which case multi-dimensional parameter analysis may give more insight than trying to break down the problem to simpler units. Furthermore, the processing method used to obtain a given solid product has obviously a very important effect on product characteristics, e.g. granules produced by spray drying, differ considerably from those produced by high shear granulation. A product in the form of a tablet can be very different from the same product produced by extrusion. Despite these complications, it should be observed that all solid products are formed from individual grains assembled together in a structure that depends on the relative sizes, shapes and quantities of the individual grains. In what follows the intention is to suggest that an important part of solid product design is particle packing theory completed by a correct approach to establishing mixture quality and stability.

2 COMPACITY OF GRANULES OR TABLETS

2.1 Binary Mixtures of Very Small and Very Big Particles $X_{big}/X_{small} > 7$

A relatively simple case to deal with is the determination of mixture proportions to give a maximum compacity (or minimum porosity as $\varepsilon = 1 - c$). This could correspond to a tablet of good mechanical strength, a granule with a high density or the optimum filling of a capsule. The reasoning is based on particle packing theory.[1,2] It is well known that the porosity of a pack of particles is independent of the absolute size of the grains, if surface effects can be neglected. Thus combining two powders with identical particle size will not affect the porosity of the mixture. However, in the case where the components have different sizes it may be that the small particles can fit in between the large particles to better fill space and reduce the porosity. It is easy to calculate the porosity (ε) of a mixture of big and small particles. Equation 1 gives the porosity when very small particles fill the free space in a pack of very big particles. Equation 2 gives the porosity when very large particles replace a certain volume of small particles.

$$\varepsilon_{mixture} = \varepsilon_{big} - \frac{(1 - \varepsilon_{big})V_{small}}{(1 - V_{small})} \tag{1}$$

That the reduction in porosity when large spheres replace a certain volume of small spheres is :

$$\varepsilon_{mixture} = \frac{\varepsilon_{small}V_{small}}{1 - \varepsilon_{small}(1 - V_{small})} \tag{2}$$

And that these two curves meet at a minimum porosity of

$$\varepsilon_{mixture} = \varepsilon_{small} \cdot \varepsilon_{big} \tag{3}$$

at a volume fraction of small grains (V_{small}) of :

$$V_{small} = \frac{\varepsilon_{big}}{\varepsilon_{big} + \dfrac{(1 - \varepsilon_{big})}{(1 - \varepsilon_{small})}} \Rightarrow \frac{\varepsilon}{1 + \varepsilon} \quad when \quad \varepsilon_{small} = \varepsilon_{big} \tag{4}$$

These results can be used to determine the minimum porosity of a mixture of very small and very large particles as a function of the porosity of the two components. Tables 1 and 2 are calculated from expressions (3) and (4). For example if the

Table 1 *Minimum Porosity of a Binary Mixture as a Function of the Porosity of the Two Components*

Porosity	$\varepsilon_{big} = 0.3$	$\varepsilon_{big} = 0.4$	$\varepsilon_{big} = 0.6$
$\varepsilon_{small} = 0.3$	0.09	0.12	0.18
$\varepsilon_{small} = 0.4$	0.12	0.16	0.24
$\varepsilon_{small} = 0.6$	0.18	0.24	0.36

Table 2 *Volume % of the Small Size Component to Give a Mixture of Minimum Porosity as a Function of the Porosity of the Two Components*

Porosity	$\varepsilon_{big} = 0.3$	$\varepsilon_{big} = 0.4$	$\varepsilon_{big} = 0.6$
$\varepsilon_{small} = 0.3$	23%	32%	51%
$\varepsilon_{small} = 0.4$	20%	29%	47%
$\varepsilon_{small} = 0.6$	15%	21%	38%

porosity of each component is 0.5 the minimum porosity of a mixture will be 0.25 and obtained for a volume fraction of small particles of 33%.

It can be seen that if the two components are available with porosities that vary from 0.3 to 0.6, then the minimum porosity of the mixture can vary from 0.09 to 0.36. The volume fraction of small particles for which this minimum porosity is obtained varies from 15% small particles to 51% small particles. Thus, depending on the characteristics of the separate components, different formula can be used to obtain a mixture with maximum compacity.

2.2 Binary Mixtures of Particles with Size Ratio $X_{big} / X_{small} < 7$

The simple calculation given above has its limits. In particular it is assumed that there are no interactions between the two components, no surface forces to organise assembly, nor size effects causing interference between the two types of particle. It has been shown[1] that the simple reasoning given above is valid for particle size ratios greater than 7. For particle size ratios below this limit it is possible to calculate structure based on a simple packing model.[2]

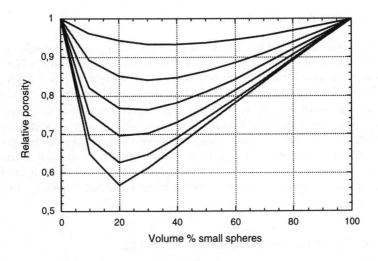

Figure 1(a) *The variation of porosity of binary mixtures of spheres as a function of the size ratio and the mixture composition*

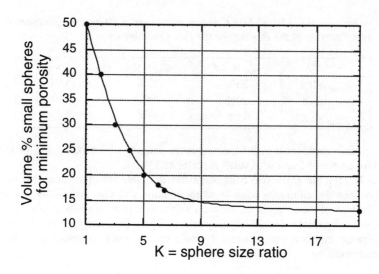

Figure 1(b) *The volume percentage of small particles for a minimum porosity binary mixture as a function of the size ratio of the two components*

Figure 1(a) shows such results for the variation of porosity of binary mixtures of spheres as a function of the size ratio and the mixture composition. These same results are shown differently on Figure 1(b), which gives the volume, percentage of small particles required to form a binary mixture with minimum porosity as a function of the size ratio of the two components. For example if the formula requires a volume percentage of say 20% small particles then a size ratio of 5 should be used to obtain a mixture with minimum porosity or maximum compacity.

2.3 Experimental Tests of Tablet Strength

These ideas have been tested by making tablets of binary mixtures of caffeine and starch and testing their crush resistance. The results of these experiments shown in Figure 2 are for tablets made from caffeine sieved between 400 and 800 µm to give a mean particle size of $X_V = 727$ µm, mixed with StaRX1500® starch sieved between 50 µm and 100 µm to give a mean particle size of $X_V = 76$ µm. The particle size ratio is thus $X_{big}/X_{small} = 10$. Approximately 1 litre lots of powder mixture were processed in a Killian 15-punch rotary tablet press using only three punches. The punch displacements and the punch pressure was fixed and not varied for each of the lots. The crush resistance of samples of the tablets made were measured with an Eweka TBH 30 hardness tester. It can be seen that the variation in strength corresponds to that of the calculated compacity ($1-\varepsilon$) shown in Figure 1(a).

3 DISPERSABILTY OF TABLETS OR GRANULES

Another case of solid product formulation where theory can give some indications is for the optimum composition for rapid dispersion tablets or granules. In such tablets a

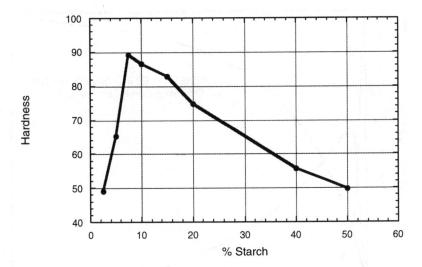

Figure 2 *Crush resistance of caffeine-starch tablets as a function of starch content*

relatively insoluble active ingredient can be associated with an easily soluble ingredient which on immersion takes up water, thus breaking down the structure of the tablet and liberating the active ingredient as a dispersion of particles.

3.1 Maximise Contacts Between the Grains

As a first approach to formulating such a product it may be assumed that the structure of the tablets or granules should be that contact is favoured between the grains of hydrophilic material and the grains of active ingredient. The formulation of the mixture can therefore be based on calculations of contacts between particles in packings. Figure 3(a) shows the results of such calculations for the case of a mixture of two sphere sizes with a size ratio of 5. This gives the percentage of contacts between small grains t_{11}, between large grains t_{55} and between dissimilar grains t_{15} where $t_{11} + t_{55} + t_{15} = 1$. It can be seen that a mixture with about 5% of small particles (size 1) by volume mixed with 92% large particles (size 5) by volume gives a maximum number of contacts between large and small particles. It may be concluded that tablets of this composition could have good dispersion. The Figure 3(b) shows the results of similar calculations for other size ratios and gives the volume percentage of small spheres which will give a maximum number of dissimilar contacts as a function of the size ratio of the component in the mixture. For example if a formula imposes 20% by volume of an active ingredient in a tablet then this should have a particle size half that of the hydrophilic particles. However if the required composition is 80% of active ingredient then the hydrophilic component should have a particle size twice as big.

3.2 Continuity of the Hydrophilic Network

However, the calculation based purely on maximising the contacts between grains does not cover the entire problem. It is also necessary for the contacts to form a continuous hydrophilic network in the tablet or granule so that water taken up is transported

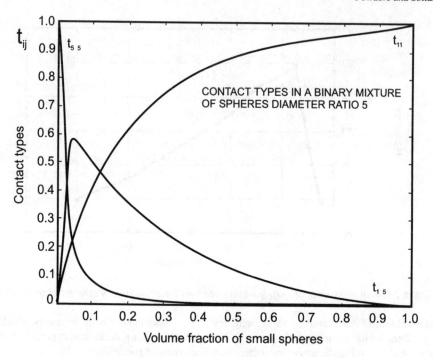

Figure 3(a) *The percentage of contacts between small grains t_{11}, between large grains t_{55} and between dissimilar grains t_{15} for a binary packing with size ratio 5*

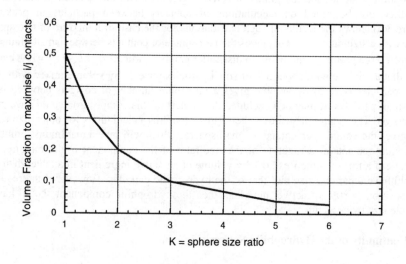

Figure 3(b) *Volume fraction of small particles to maximise contacts between the two components in a binary mixture as a function of particle size ratio*

throughout the tablet to ensure break up. A general approach to problems of path continuity in networks is by the theory of percolation which was first proposed in a study of progressive blocking of filters and has since been developed as a rigorous mathematical theory.[3]

An application of the theory, which has value as an example, is in following the variation in overall electrical conductivity of a mixture of conducting and non-conducting particles as a function of the content of conducting particles. Clearly the first few conducting particles added to a set of non-conducting particles will have no effect on the overall conductivity. Only when sufficient conducting particles are added to give a continuous conducting network from one end of the bed to the other will the overall conductivity be different from zero. The composition where conduction starts is reproducible, well identified composition called the percolation threshold. It corresponds to when an infinite cluster of conducting particles co-exists with the non-conducting particles. When the amount of conducting particle is less than the percolation threshold there are only finite size clusters of conducting particles that do not give an overall electrical conductivity. The mean size of the finite clusters, the exact value of the percolation threshold, the way the conductivity varies with the percentage of conducting particle all depend on the network characteristics and can be predicted by percolation theory. In this example of mixing equal sized spheres the percolation threshold is known to be about 29% conducting spheres corresponding to a contact network with a coordination of about 6.

This same theory should be applicable to the problem of defining the relative amounts hydrophilic particles to be added to some insoluble particles and create a continuous hydrophilic network in the mixture. However, the example given above, well understood and abundantly commented in the literature, is not directly applicable to the case where the two components are different particle sizes. For example in a mixture with size ratio 10 the large particles can have a contact coordinance with about 380 small particles. In turn the small particles can have a contact coordinance with the large particles of the order of 4. Clearly the contact network is more complicated than in the case for mono-sized particles and consequently the value of the percolation threshold will be very different. The problem of the percolation threshold in binary mixtures has been investigated by means of computer simulations and electrical conductivity measurements.[4] The results are summarised in Figures 4 and 5. It can be seen that the results for a size ratio of 10 are outside the range of sizes considered but could correspond to a percolation threshold of 10%. The minimum percolation threshold is at a size ratio of about 5.

3.3 Measurements of the Disintegration Time of Caffeine-starch Tablets

Tablets made from caffeine and starch mixtures as described in Section 2 were tested for disintegration time using an Eweka ZT 32 disintegration tester. The results are shown in Figure 6 where it can be seen that at about 10% starch gives a minimum disintegration time which does not change significantly for higher starch contents. Further experiments are required to decide whether this result should be interpreted as either a percolation threshold or as a maximum in the contacts between dissimilar particles.

4 MIXING POWDERS

Once the mixture proportions of a solid product are decided it is necessary to mix the

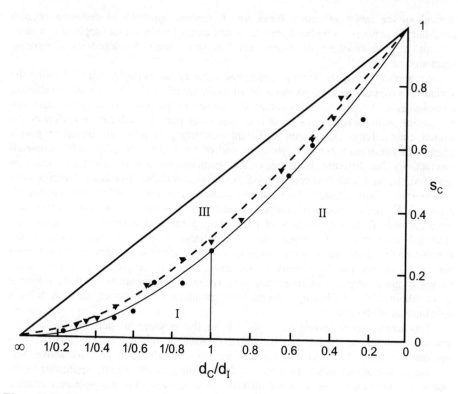

Figure 4 *Number fraction percolation threshold in binary mixture of conducting and non-conducting spheres with different sizes.* • *Experimental,* ▼ *Numerical simulations.[4]*

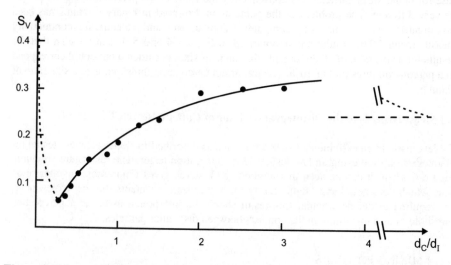

Figure 5 *Volume fraction percolation threshold in binary mixture of conducting and non-conducting spheres with different sizes.[4]* • *Experimental*

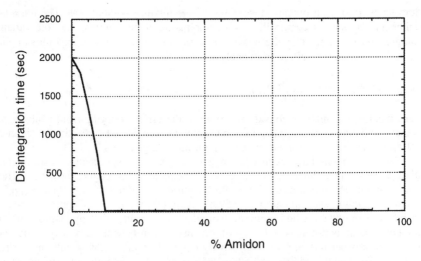

Figure 6 *Disintegration time and starch content in binary tablets*

ingredients together. Powder mixing is thus fundamental to solids formulation but the practical approach to this operation remains mainly empirical. There are two main distinguishing features of solids mixing. Firstly the mixture is made at the level of individual particles and not at molecular level thus powder mixtures are always intrinsically heterogeneous. Secondly molecular diffusion is not operative and mixing can only be made by mechanically moving particles with respect to each other thus potentially introducing segregation effects. The complexity of interactions between particles (particle size and shape, density, cohesion,.....), and the present state of knowledge of segregation phenomena as a function of the characteristics of mixers (geometry, operating conditions, methods of filling and emptying, ..;) does not allow anything other than trial and error design.

Apart from the important practical decision on how to perform the mixing of powder ingredients it is also necessary to examine how to evaluate mixture quality and the stability of the powder mixture in subsequent handling. If the powders are not well mixed or tend to de-mix then the theoretical considerations on the respective roles of the different ingredient mentioned above, come to nothing.

4.1 Evaluating the Quality of a Mixture of Powders

The degree of mixing and lot uniformity are important quantities affecting product quality which can not be dissociated from the problem of sampling. In effect a mixture is defined as being homogeneous if a series of samples have the same composition, and the same properties. The number of samples, the sample size, and the location where they are taken are fundamental for characterising mixture quality.

Sample size is intimately linked to the final use of the powder mixture. In the problems treated here it could be the size of a tablet or a granule. Furthermore the sample size must be suited to the particle size distribution so as to avoid a size selection phenomenon. The number of samples should not be too great, as taking each sample can perturb the mixture and affect subsequent samples, but sufficient should be taken to be statistically representative. Finally the location of each sample should be chosen at random so as to give an overall view of the whole mixture without bias. The measured experimental variance of composition is a combination of three terms : the true variance in composition resulting from the mixing process, the variance introduced by sampling errors and the variance resulting from analysis.

$$\sigma^2_{measured} = \sigma^2_{mixture} + \sigma^2_{sampling} + \sigma^2_{analysis} \tag{5}$$

Modern methods of analysis considerably reduce the errors resulting from analysis. On the other hand the sampling operation must not be considered as a simple technical operation but as a random process which can introduce large errors.[5]

These problems have been examined in a series of experiments[6,7] using a high dilution pharmaceutical binary mixture of two components with roughly the same particle size distribution. The mixtures used were 99% lactose (X_{10}=20μm, X_{50}= 100μm, X_{90} = 225 μm) and 1% sodium saccharinate (X_{10}=59μm, X_{50}= 115μm, X_{90} = 197 μm). In order to examine the sampling method a mixture of 75g was "handmade" by introducing the minor component between two layers of the major component in a bottle and then shaking. The powder mixture was then slowly emptied on to a moving belt and entirely collected as 510 samples of 130 mg each called unit size X=1. Sample sizes greater than 1 can be taken into account by summing with neighbouring samples. The saccharinate content of each sample was determined by dilution in 50ml buffer solution a pH 8 and analysed by HPLC. This is illustrated in Figures 7(a) and 7(b).

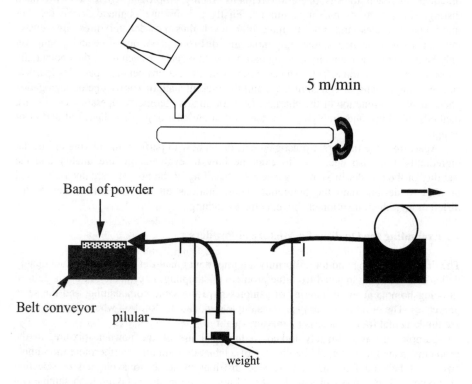

Figure 7(a) *Sampling methodology*

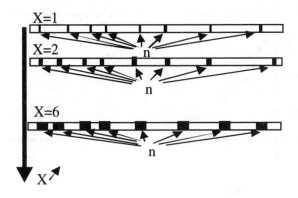

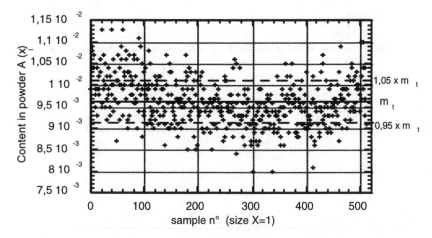

Figure 7(b) *Description of sample size X, with results of the 510 sample compositions*

The results of these experiments are shown in Figures 8 and 9. From Figure 7(b) the "true" mean and "true" standard deviation of the saccharinate content can be calculated. From Figure 8 it can be seen that as the number of samples considered increases the extreme values of the results are centred around the "true" mean . It may be concluded from this figure that for a number of samples greater than 5 the extreme values are equal to the "true" mean ± 5%. A similar reasoning can be made for the standard deviation and shows that in this case the number of samples for a 95% confidence should be equal or greater than 10. Finally Figure 9 shows the influence of the sample size on the "true" standard deviation of saccharinate content and indicates that a sample size of 10 is adequate for calculating mixture content. This corresponds to the knee of the curve the form of which indicates that there is a structure in the mixture when spread along the moving belt.

4.2 Structure of a Mixture

However, the evaluation of the mixture quality described above may not be pertinent in

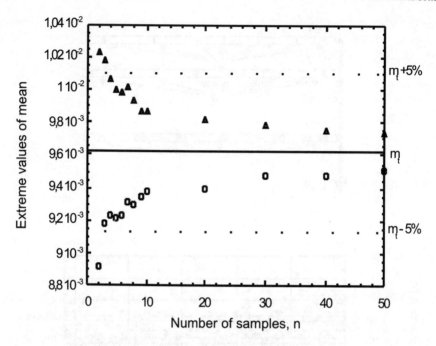

Figure 8 *Extreme values of the mean for the different populations studied (X = 1)*

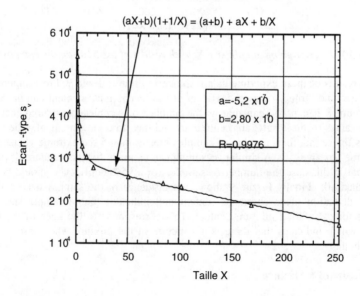

Figure 9 *Effect of sample size*

the formulation cases described above. It is necessary to take into account the fact that the product structure is based on having particles with very different sizes which implies having very great differences in mass and number compositions. In addition as the required properties depend on the arrangements of contacts between particles then the geometry of contacts must be taken in to account in the mixture quality. A method of doing this based on coordination numbers has been proposed,[8] but has not yet found wide acceptance.

5 CONCLUSION

In the simples cases treated here where formulations are binary mixtures of particles of different sizes, it is obvious that the role of the small particles in the mixture is different from that of the large particles. The critical volume fraction for whatever property is important in a tablet (hardness, dispersion, ..) are always found for mixtures with smaller proportions by volume of small particle than large and usually in the range of volume fractions less than 30%. This should be borne in mind when deciding on the powders to be used in a given case. For example, if the active ingredient is the minor component in a tablet or granule then it should be chosen as the finer powder. Whilst this conclusion is rather obvious, if only from the practical point of view of better ensuring a uniform distribution of active ingredient in a tablet, it also suggests that the less intuitive conclusions may also be reasonable. When the particle sizes of the components and optimum proportions of the formula are determined it remains to make the practical mixture. This requires moving the particles with respect to one another to achieve an intimate mix such that the large and small particle contact with one another in the required fashion, thus posing technological problems in making the mixture with a potentially segregating mixture and also in evaluating the quality of the mixture.

References

1. R. M. German, *Particle Packing Characteristics*, Metal Pow. Ind. Fed. Princeton, 1984.
2. P. Troadec and J. A. Dodds, Chapter 8 in *Disorder and Granular Media,* Eds. D. Bideau and A. Hansen, Elsevier, Amsterdam, 1993
3. H. Leuenberger, *Advanced Powder Technology*, 1999, **10**, 323-352.
4. L. Oger, J. P. Troadec, D. Bideau, J. A. Dodds and M. J. Powell, *Powder Technology*, 1986, **46,** 121-131, 133-140.
5. P. Gy, *Sampling of Particulate Materials*, 2nd Edition, Elsevier, Amsterdam,1982.
6. S. Chaudeur, Doctorate Thesis, INPL, Nancy 2000.
7. S. Chaudeur and H. Berthiaux (in preparation).
8. L. T. Fan, J. R. Too, F. S. Lai and Y. Akao, *Powder Technology*, 1979, **22**, 205.

GRANULAR POWDERS AND SOLIDS: INSIGHTS FROM NUMERICAL SIMULATIONS

M. Ghadiri, S. J. Antony, R. Moreno and Z. Ning

Department of Chemical and Process Engineering
University of Surrey
Guildford GU2 7XH
U.K.

1 INTRODUCTION

Granular materials are an important part of many industrial processes. In a wide range of applications, it is often necessary to disintegrate and disperse particles in gasses or liquids under shear flow. Interestingly, recent investigations[1] for hydrodynamic interactions on concentrated systems reveal that the distribution of stresses in flowing aggregated suspensions exhibits similarities with the stress distribution in granular systems. An estimate of the shear resistance of such a particulate medium is of great importance to the process control. Current practice is based on empirical approaches for which it is difficult to specify the limits of operation. This often leads to production problems and inefficient or ineffective equipment design. Understanding to control the behaviour of granular materials requires understanding of the physical process that controls the behaviour and interaction of their constituent particles. This requires understanding and modelling of the ways in which they flow, the ways in which groups of particles form, different sizes segregate and individual particles fracture or wear out. For example, in rheological problems, a theoretical analysis (based on continuum theories) for flow fields of aggregated concentrated suspensions under slow shear flow seems to be applicable if the size of suspension particles is an order of magnitude higher than that of fluid particles. Although it is customary to treat the size of suspension an order of magnitude higher than that of fluid particle, to the best of our knowledge, no precise reason is available in the literature to justify this. This requires detailed investigations on how the size effects of particles could influence the flow behaviour.

In this paper, we review some of the research carried by the authors since 1997 on (a) shear characteristics of granular materials including attrition process and (b) breakage process of spherical agglomerates. The simulations were carried out using the Discrete Element Method (DEM), which was originally developed by Cundall and Strack.[2] The method models the interaction between contiguous particles as a dynamic process and the time evolution of the particles is advanced using an explicit finite difference scheme. The interaction between the neighbouring particles is modelled by algorithms based on theoretical contact mechanics provided by Thornton and Yin[3] and Thornton.[4] For single particle impacts, the simulation results have been confirmed by matching analytical solutions when available, as reported by Thornton and Yin[3] and Ning.[5]

2 SIZE EFFECTS IN SHEARED PARTICULATE MEDIA

Previous knowledge on the size effects (three-dimensional) in a granular assembly subjected to shearing, as encountered in real practice, is scarce in the literature. However, using an analytical and experimental approach, Gundepudi et al.[6] have studied the stress distribution inside a single spherical particle when supported by multiple contacts. They have observed that as the number of contact points increases, the maximum internal tensile stress (responsible for internal fracture), in general, decreases as the stress state inside the sphere approaches hydrostatic compression. An increase in the number of contacts supporting the sphere helps to increase the hydrostatic (mean) stress in the sphere. Recently, Tsoungui *et al.*[7] have reported two-dimensional numerical (Molecular Dynamics Simulations) and experimental results (Figure 1) for the crushing of grains inside a two dimensional polydispersed granular assembly subjected to oedometric compression. The stress state of the grains has been decomposed into two components; namely, the hydrostatic part and the deviatoric part. They have observed that the grain fragmentation is difficult when the size of the larger particle (surrounded by smaller particles) increases. They have argued that the particles surrounding larger particle produces an increase in hydrostatic stress and a decrease in deviator stress of the larger particle. It is interesting to note from the experiments that, despite an increase in external pressure, the larger grain fracture became impossible due to this hydrostatic effect. Very recently, Nakata *et al.*[8] have reported some interesting experimental and numerical results (three-dimensional, using DEM) for the bulk crushing strength of granular assembly. They have observed that (i) the crushing strength decreases with an increase in particle size (ii) the grain split probability increases with the applied stress and the number of flaws and (iii) the grain split probability decreases with an increase in number of particle contacts and particle size. They have attributed these findings to the possible increase in mean stress of the grain when they are surrounded by many contacts.

In general, the analysis of these trends using an experimental approach is difficult, as any attempt to measure the quantities inside the granular media would influence its behaviour. Moreover the experimental measurements contain a large degree of error. For example, the measurement of contact forces which play an important role in mobilising the shear resistance,[9] is quite difficult. In these situations, performing numerical

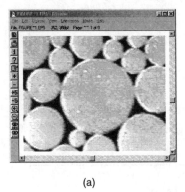 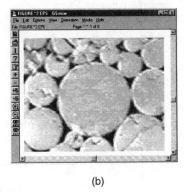

(a)   (b)

Figure 1 (a) *Enlargement of a region of granular material showing the hydrostatic effect on a large grain surrounded with small grains (b) Despite the increase of external pressure, the large grain fracture becomes difficult because of this hydrostatic pressure*[7]

simulations using DEM would be an ideal choice. In this section, we analyse the size effects in granular media subjected to slow shearing using DEM. The simulations are performed in a periodic cell in which a large size particle (submerged particle) is created at the centre of the cell and surrounded by 5000 randomly generated mono-dispersed spherical particles. The boundaries of the periodic cell from the centre of the cell were at a distance of c.a. 5 times radius of the submerged particle. Different values of the size ratio (ratio of the diameter of submerged particle to that of surrounding particles) were considered, viz., 5, 7.5, 10 and 15.

All the samples considered here had elastic properties corresponding to 'hard' particles (Young's modulus E = 70 GPa   Poisson's ratio = 0.3, coefficient of interparticle friction = 0.3, and interface energy = 0.6 J m^{-2}). After the particles were initially generated, a servo-control algorithm was used to isotropically compress until a mean stress p = 100 kPa was achieved. At the end of the isotropic compression, the microstructure of the samples was isotropic. At this stage, the solid fraction and mechanical coordination number (average number of load bearing contacts) of the samples considered in this study were 0.650±0.017 and 5.83±0.26, respectively. For shearing, a strain rate of 10^{-5} s^{-1} was employed in the simulations. The samples were subjected to the axi-symmetric compression test ($\sigma_1 > \sigma_2 = \sigma_3$). During shearing, the mean stress p = ($\sigma_1 + \sigma_2 + \sigma_3$)/3 was maintained constant at 100 kPa using the servo-control algorithm.

Figure 2 shows the ratio of principal deviatoric stress component (τ_{Ds}) to the hydrostatic stress component (p_s) of the submerged particle at the steady state during shearing. For an increase in the size ratio, τ_{Ds}/p_s decreases, thereby, showing a growing dominance of the hydrostatic component of the submerged particle with the size ratio. A detailed mathematical analysis of this has been reported elsewhere.[10] The nature of stress is fully dominated by the hydrostatic stress component for particles with a size ratio equal to or greater than c 10. Therefore, for larger particles (with size ratio equal to or greater than c 10) in a dense system undergoing shearing, the particles would have less resistance

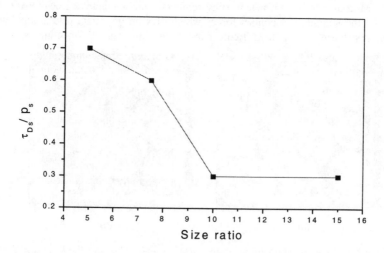

Figure 2 *Variation of the ratio of principal deviatoric stress component to the hydrostatic stress component with size ratio of submerged particle at steady state*

to move. This observation is relevant to the segregation phenomena occurring in many industrial processes.

3 ATTRITION AND FRAGMENTATION OF PARTICULATE SOLIDS

The shear cell test is widely used for measuring the attrition rate as well as the mechanical properties of bulk particulate solids. However, almost all the work in this area is based on experimental approaches and there is little theoretical analysis of the process. The reported investigations on the degradation of particulate materials are mainly descriptive and lack direct quantitative verification. Fundamental information on attrition in a shear cell, such as packing geometry, internal force transmission, and velocity distributions are still outside the scope of current experimental techniques. The distinct element method (DEM) is able to provide fundamental information on the particle system, such as evolution of particle locations, interparticle forces and particle velocities. For the shear cell simulation, the particle motion caused by shearing and the particle breakage due to the generation of high force chains can be monitored. Most importantly, the contributions to attrition by surface wear and particle fragmentation can be de-coupled, a feature that is difficult to achieve by traditional experimental approaches.

Ghadiri *et al.*[11] examined various models of attrition of granular solids in an annular shear cell reported in the literature. They showed that the models proposed by Neil and Bridgwater[12] and Ouwerkerk[13] unified the experimental data satisfactorily for the formation of fine debris, but when the coarse fragments and non-spherical particles were included, the trend of the experimental data did not follow the models satisfactorily. By examining the shape and origin of the debris, Ghadiri *et al.*[11] concluded that the model of Neil and Bridgwater applied well to surface damage processes, such as wear, but not to the fragmentation process. Furthermore, the influence of the mechanical properties on the two processes of wear and fragmentation under shear deformation could not easily be described in a general way by a single correlation. In this paper, the process of attrition in a shear cell is analysed by computer simulations, using the distinct element method, where appropriate models of the mechanisms of surface wear and fragmentation have been incorporated into the computer code. The side crushing strength (SCS) of single particles is used as the criterion for fragmentation. For surface damage processes, the abrasive wear by the formation of lateral cracks during shearing is calculated using a semi-brittle failure model proposed by Ghadiri and Zhang.[14] The contributions to attrition by surface wear and particle fragmentation are therefore separated. The results are compared with the experimental data reported by Ghadiri *et al.*[11]

The particles of interest here are spherical catalyst carrier beads. A periodic shear cell sample consisting of 700 primary particles was prepared for the computer simulation of the shear cell tests as shown in Figure 3(a). The annular shear cell used in the experiments had saw-tooth grooved surfaces in order to prevent particle slipping at the shearing plates. In the computer simulation, this situation was approximated using rough particle planes. As shown in Figure 3(b), the sample particles at the bottom and top planes (shown as black particles) are specified not to move with respect to each other and are all given a velocity along the x-axis, which is the shearing direction. For the boundary particles, there is no movement in the z direction, but motion in the y direction is allowed in order to maintain the normal stress level.

A servomechanism is used to maintain constant the normal component of the resultant force on the rough particle wall. The servo-control continuously adjusts the velocity of the wall such that the difference between the calculated normal stress of the

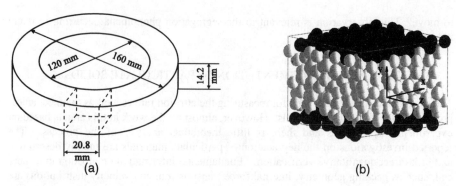

Figure 3 *Annular shear cell and the generated assembly: (a) a segment in an annular shear cell; (b) a periodic cell of particles prepared for simulation*

wall and the required stress is minimised. For the details of servomechanism and implementation into the computer code, readers could refer to Cundall and Strack.[2] As shown in Figures 3(b), the two boundary planes that are perpendicular to the *x*-axis are periodic planes, i.e. once a particle is out of the cell in one plane it will immediately join in the opposite plane. Periodic boundary control also applies to the two planes perpendicular to the *z*-axis.

The procedures of sample preparation are briefly described as follows. Within a given cubic rectangle region, primary particles are randomly generated. The assembly is then compacted by applying strain rates in the *x*, *y* and *z* directions. At the same time a normal stress tensor parallel to the *y* direction is specified with a magnitude equal to the applied load divided by the shearing plate area. While the size of the periodic cell decreases due to compaction, the number of contacts gradually increases, and the density or solid fraction of the cell assembly also increases. When the co-ordination number, defined as the average contact number for each individual particles, or solid fraction of the assembly, reaches the desired level, interfacial friction is introduced in very small increments. Since the gravity force is small compared to the contact forces generated by the external normal stress, the effect of gravity is not considered in this study. Therefore, the gravity value is set to zero for all the particles. The parameters of the generated assembly are shown in Table 1.

The particle properties used in the simulations are reported by Ghadiri *et al.*[11] and are summarised in Table 1. One of the important properties determining the extent of attrition in a shear cell is the Side Crushing Strength (SCS). The distribution of the SCS for the measured assembly is shown in Figure 4 with an average value of 44.9 N. Each particle within the assembly is assigned with a certain value of the side crushing strength

Table 1 *Particle Properties and Assembly Parameters Used for Simulation*

	Catalyst beads		Assembly parameters
Young's modulus *(GPa)*	31.0	Particle number	700
Poisson ratio	0.2	Co-ordination number	4.5
Density *(kg m⁻³)*	800	Mean particle diameter *(mm)*	2.2
Yield stress *(GPa)*	0.25	Periodic cell size *(mm³)*	22.8×14.2×20.0
Fracture toughness *(MN m⁻³ᐟ²)*	0.2	Solid fraction	0.58
Coefficient of friction	0.4		

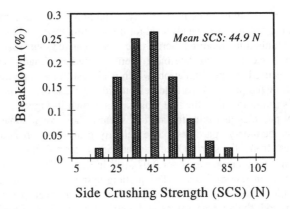

Figure 4 *Side crushing strength (SCS) of silica catalyst carrier beads (from Kenter[15])*

such that the distribution of the 'strength' within the assembly is the same as shown in Figure 4. The simulated assembly is a polydispersed system containing three different particle sizes. The particle radii and corresponding mass percentages are 1.05 mm, 25%;

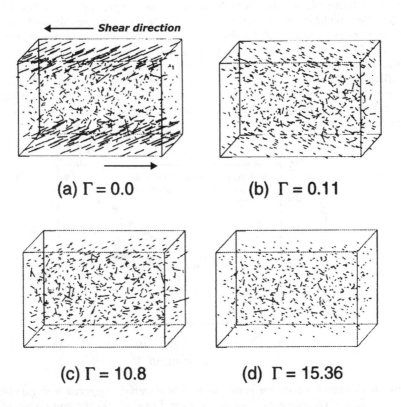

Figure 5 *Velocity distribution in the assembly at different shear strain levels Γ. The magnitude of the velocity is proportional to the line length*

1.1 mm, 50%; and 1.15 mm, 25%, respectively. The above choice is based on the size distribution of the catalyst carrier beads used for the attrition tests.

Computer simulations reported here have been carried out for normal stresses 100 *kPa*, for which experimental data have been reported by Ghadiri *et al.*[11] A shear strain rate 7.0 *s*$^{-1}$ is used for the simulation, which is closest to the experimental condition. Figures 5(a-d) show the profile of particle velocity in the shear cell, where the magnitude of the velocity is proportional to the line length. As shown in Figure 5(a), the velocities of non-boundary particles just after the start of shearing are very small as compared with the static boundary particles. The velocity profile shown here is after a few calculation cycles from the start. Once the shearing is started, the average velocity of non-boundary particles increases to about 0.07 m s^{-1} in a short time (equivalent to Γ= 0.1-0.2) at the shear strain rate of 7.0 s^{-1}. The particle system is transformed from a static state to a dynamic state. For a given normal load and a shear strain rate, after this transformation period, the average values and distribution profiles of both contact forces and particle velocities remain roughly constant during the whole process of shearing. Figure 6 shows the results of the simulations of the attrition rate as a function of shear strain under the applied normal load of 100 kPa. The experimental data of Ghadiri *et al.*[11] are also incorporated into the figure for comparison. The contributions to attrition by surface wear and particle fragmentation have been separated. The results show that, for the material used here and the applied normal load of 100 *kPa*, body fragmentation plays a more dominant role on attrition in the shear cell as compared to the surface wear. In general, the simulation results agree with experimental data reasonably well. However, there is a visible difference of the slopes between the two attrition curves, notably when Γ > 10. In the experiments, the processes of surface wear and fragmentation produce debris, which modify the force fabric. For small shear strains, this effect is not significant. However, for large shear strains, the presence of debris reduces the bed voidage, leading to an increase in the number of interparticle contacts. Hence, the contact forces could be re-distributed more homogeneously, resulting in a reduction in the rate of attrition.

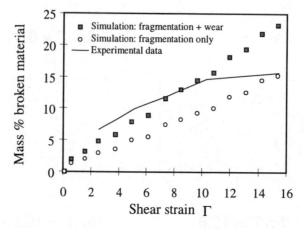

Figure 6 *Computer simulation results of broken material compared with experimental data of Ghadiri et al.*[11] *under the normal load of 100 kPa and the shear strain rate of 7.0 s*$^{-1}$

4 IMPACT AND FRAGMENTATION OF AGGLOMERATES

In several industrial applications, the granular particles are bonded together to form agglomerates. This is done, for example, to enhance flow and transportation operations or to facilitate a controlled dispersion process. Information on the strength of agglomerates are vital as they depend on the properties of constituent particles, packing, the way agglomerates are created and the mechanical and environmental conditions under which the agglomerates operate. Previous studies on the impact behaviour of agglomerates were performed in which particles of different sizes were randomly arranged to form the agglomerates.[16] In this paper, we perform numerical simulations (using DEM) for agglomerates having monodispersed distribution of particles, with an uniform internal structure. Different values of surface energy of constituent particles have been attributed to the agglomerates and tested with a range of impact velocities. We show that the failure mode of agglomerates can vary from disintegration to fragmentation by simply varying the adhesion properties of the particles, even for the agglomerate with low solid fraction.

The agglomerates were created using 3000 monodispersed particles with the following properties: Young's modulus E = 31 GPa, Poisson's ratio = 0.2, coefficient of interparticle friction = 0.35, and interface energy γ = 3.5 J/m^2 (weak agglomerate), 35 J/m^2 (strong agglomerate), density = 2000 kg/m^3. The primary particles were initially generated within a spherical region and then brought together by introducing a centripetal gravity field. The diameter of the agglomerate thus generated was 0.907 mm, the solid fraction was 0.546 and the coordination number was 5.62 . At this stage, the centripetal field introduced earlier was switched off. It was observed that the contact distribution of the particles within the agglomerate was uniform. This was confirmed by obtaining the same value of the average number of contacts, present at different annular segments from the centre of agglomerate. For simulating the impact process of agglomerate, a rigid wall was created close to the bottom of the agglomerate. The agglomerate was impacted (in the presence of gravity) against a wall at different velocities and the dependence of breakage process on velocity was identified.

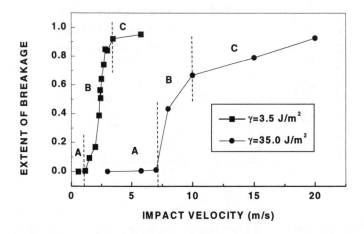

Figure 7 *Extent of breakage after impact as a function of the impact velocity*

Figure 7 shows the extent of breakage of the agglomerate as a function of impact velocity for the strong and weak agglomerates. The extent of breakage has been identified[17] in terms of the ratio of mass of the clusters produced (excluding the largest cluster) to the initial mass of the agglomerate. As expected, the simulations show that the agglomerate with high interface energy has higher strength. The extent of breakage of agglomerates, in general, can be represented in three different regimes A-C (Fig.7) in terms of the impact velocity. Figure 8 shows the size distribution of the two largest clusters in these regimes. Figure 9 shows the typical visual presentation for the fragmentation of agglomerates. The agglomerates, after impact, have been viewed at different positions (views 1-4, & top and bottom view).

Regime A: In this regime, chipping was observed and the agglomerates rebounded after the impact. For the weak agglomerate, this regime occurred at an impact velocity between 0.5-1.5 m/s and for the strong agglomerate, this regime spans between 5-7.5 m/s. For impact velocities lower than the tail of this regime, no chipping of particles from the

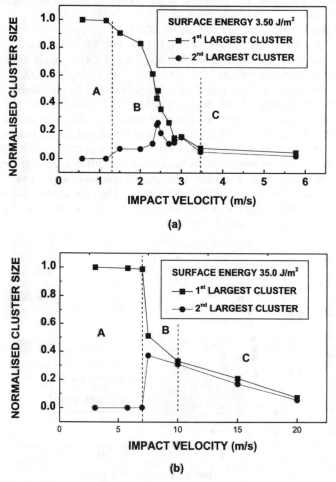

Figure 8 *Size distribution of the two largest clusters after impact: (a) weak agglomerate and (b) strong agglomerate*

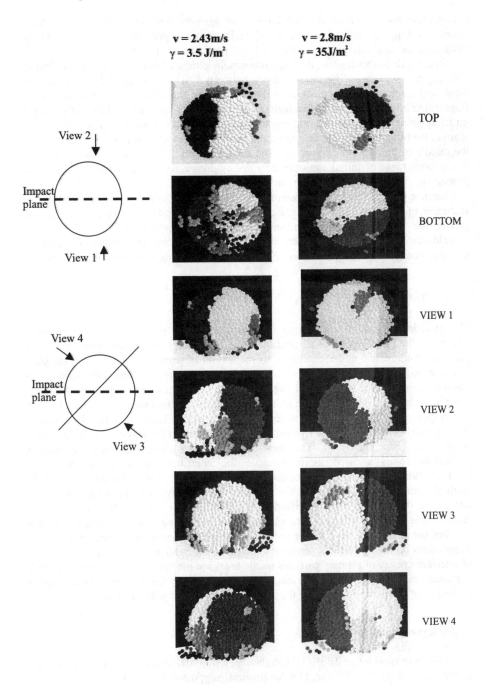

Figure 9 *Visual observations of the fragmentation of the agglomerate for different values of the surface energy.*

agglomerate was observed and full rebound of agglomerate was observed. The mass detached (chipping mode) from the initial agglomerate, in general, was about 2% of the mass of initial agglomerate (size between 1-200 particles, refer Figure 8).

Regime B: In this regime, the agglomerates, in general, failed in fragmentation mode. This regime occurred at impact velocities between 1.5-3.4 m/s and 7.5-10 m/s for the weak and strong agglomerates, respectively. In this regime, the strong agglomerates fragmented into nearly two halves, with no substantial production of debris (90% mass of initial agglomerate fragmented into two halves). In the case of weak agglomerates in this regime, the transition from the chipping mode to the fragmentation mode occurs nearly at the middle of this regime (Figure 8). The fragmentation in weak agglomerates occurred at c.a. 66% mass of initial agglomerate. For agglomerates failed in this regime, crack propagation was observed from the impact site.

Regime C: In this velocity regime, the agglomerate failure was dominated by the shattering of particles. This regime occurred at impact velocities greater than 3.4 m/s and greater than 10 m/s for the weak and strong agglomerates, respectively.

Hence, the simulations presented here clearly indicate the strong influence of interface energy of primary particles on the breakage mechanism of the agglomerates.

5 CONCLUSIONS

The particle technology community requires answers to problems of process design relating to the manufacture of particulate solids. Computer simulations presented above provided fundamental understanding of the behaviour of particulate solid on (a) the shear characteristics of granular materials including the attrition process and (b) the influence of surface energy on the breakage of agglomerates. The results presented above show a promising trend in analysing these industrial problems by the Discrete Element Method. Investigations on the size effects in sheared particulate media reveals that the nature of stress is fully dominated by the hydrostatic stress component for particles with a size ratio equal to or greater than c 10. The shear cell simulations carried out for the attrition of particulate solids allowed us to separate the contributions of surface wear and fragmentation of particles. The results show that, for the material used here and the applied normal load of 100 *kPa*, body fragmentation plays a more dominant role on attrition in the shear cell as compared to the surface wear. In general, the simulation results agree with experimental data reasonably well especially for shear strains less than 10. Further analysis is required for improving the predictions at large shear strain levels. Simulations on the impact characteristics of agglomerates have shown a strong influence of interface energy of primary particles on the breakage mechanism. The failure mode of agglomerates varied from disintegration to fragmentation by simply varying the adhesion properties of the particles, even for the agglomerate with low solid fraction.

Acknowledgements

The authors acknowledge EPSRC (Grant No. GR/M33907) and ICI Strategic Technology Group Technology Ltd., Wilton, U.K for financial support of this work.

References

1. L. E. Silbert, R. S. Farr, J. R. Melrose and R. C. Ball, Jl. Chem. Phys., 1999, **111** (**10**), 4780-4789.
2. P. A. Cundal and O. D. L. Strack, Geotechnique, 1979, **29**, 47-65.
3. C. Thornton and K. K. Yin, Powder Tech.,1991, **65**, 153-166.
4. C. Thornton, Jl. Appl. Mech., 1997, **64**, 383-386.
5. Z. Ning, PhD Thesis, 1995, Aston University, Birmingham, U.K.
6. M. K. Gundepudi, B. V. Sankar, J. J. Mecholsky, and D. C. Clupper, Powder Technology, 1997, **94**, 153-161.
7. O. Tsoungui, D. Vallet, and J. C. Charmet, Powder Technology, 1999, **105**, 190-198.
8. Y. Nakata, A. F. L. Hyde, M. Hyodo and H. Murata, Geotechnique, 1999, **49**, 567 – 583.
9. S. J. Antony, Physical Review E, American Physical Society, 2001, **63**, 011302.
10. S. J. Antony and M. Ghadiri, Jl. Applied Mechanics, ASME (Submitted).
11. M. Ghadiri, Z. Ning, S. J. Kenter and E. Puik, Chemical Engineering Science, 2000, **55**, 5445-5456.
12. A. U. Neil and J. Bridgwater, Powder Technology, 1994, **80**, 207-219.
13. C. E. D. Ouwerkerk, Powder Technology, 1991, **65**, 125.
14. M. Ghadiri and Z. Zhang, 1992, IFPRI Final Report, ARR 16-03, University of Surrey, U.K.
15. S. J. Kenter, Diploma Thesis, 1992, University of Twente, The Netherlands.
16. J. Subero, Z. Ning, M. Ghadiri and C. Thornton, Powder Technology, 1999, **105**, 66-73.
17. Z. Ning, R. Boerefijn, M. Ghadiri and C. Thornton, Adv. Powder Tech., 1997, **8(1)**, 15-37.

FLOW AID TECHNOLOGY

N. Harnby

Department of Chemical Engineering
The University of Bradford
Bradford BD7 1DP

1 INTRODUCTION

A flow aid is a minor powder ingredient which when added to a bulk powder will result in both improved interparticulate and bulk flow characteristics.

The term 'flow aid' is now common usage but specific process industries have used terms such as flow additive, flow promoter, glidant, flow conditioner, anti-caking agent, anti-adherant and anti-adhesive agent.

The terms flow aid and lubricant are commonly interchanged but a lubricant powder quite specifically reduces the coefficient of friction at contacting points to improve flow. An interlocking particle system can therefore be made more free-flowing if the lubricant can be coated on to the contacting surface. A lubricant powder can act as a flow aid but more commonly would be used to reduce particle-container friction and wear such as is found between the powder and the die wall in a tabletting process. In some cases a single additive can act as both flow aid and lubricant though it has been noted[1] that 'in general materials that are good flow aids are poor lubricants' and that 'a blend of two or more materials may be necessary to satisfy the process requirements of lubrication, flow promotion and anti-adherance'.

The addition of a flow aid represents both an additional raw material and processing cost and has generally occurred only in response to a consumer demand. Identifiable surges in use can be associated with the growth of large scale powder processes developed in the 1960s,[2-3] the introduction of high speed tabletting processes in the 1970s[4-5] and powder form fast-foods in the 1980s.[6-7] The 1990s[8-11] have been marked by the diversification of the markets in which flow aids are used and emphasis on using the aid not only to promote flow but also to take advantage of the chemical and physical properties of the aid to give add-on product and process advantages. The range of flow aids has changed little over the corresponding period of time. With cost often a major influence on selection, it is not surprising to find many 'generic' materials such as fly ash, talc, starch and clays in use. Manufactured flow aids including the metallic stearates, silica in a variety of forms, aluminium oxide, phosphates, amides, polysacharides and polyethyleneglycol can be purchased in a range of forms and grades. The manufacture of 'fumed' silicas by the high temperature vapour phase hydrolysis of silicon tetrachloride in a hydrogen, oxygen flame to produce primary particles, typically having a length of 0.2 to 0.3µm, has had a major impact on the flow aid market.

Perhaps the decade of the 1990s has seen the maturing of flow aid selection and application from the earlier shot-gun marriage procedure of 'try it' to a more scientifically reasoned selection process. The present paper looks to highlight the strengths and weaknesses in the present knowledge base and work towards a logical aid selection procedure.

2 FACTORS INFLUENCING FLOW AID CHOICE

Figure 1 indicates the principle information inputs required to effectively select, test and apply a flow aid to a new product. The public availability of information under the eight headings of Figure 1 is variable and is reflected in the literature available.

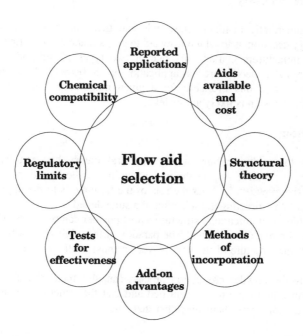

Figure 1 *Interactions in flow aid selection*

2.1 Reported Applications

The application of flow aids to specific host powders is extensively reported in the literature[1-11] and such references combined with previous company experience can provide useful historic guidance to suitable host/aid combinations. It is unlikely that an exact match for a new host powder will be found and even then the reference will not necessarily have reviewed alternative aids.

2.2 Chemical Compatibility

The chemical compatibility of the aid with the product and then with the product application is very specific and best approached by the speciality chemists of a company.

The influence of moisture and temperature on possible chemical reactions or recrystallisation must be considered.

2.3 Regulatory Limits

Limits are variable on both a national and a process basis but are very specific for food and drug products. Typically, the United States Food and Drug Administration allow fumed silica use at up to 2 percentage by weight in food preparations, up to 3 percentage by weight in cosmetic and internal pharmaceutical applications and up to 8 percentage by weight in pharmaceutical topical applications.

2.4 Aids Available and Costs

Some of the commercially available flow aids and their manufacturers are listed in Table 1. A primary decision as to which flow aid to use is made more difficult by the fact that most aid manufacturers market only one or two types of aid and a generic aid selection procedure for a specific process application would be helpful. Table 1 does not list all the product variations available for a given manufacturer and flow aid so a secondary selection decision is likely at this level.

2.5 Structural Theory

Developments in the understanding of the structural theory between particles have provided a major quantitative step in the understanding of the boundary between free-flowing and cohesive powders.[12-16] As the role of the flow aid is to disrupt structure this theory is fundamental in directing the selection of a suitable aid.

The forces which will potentially structure a host powder can be classified in terms of their magnitude and range of attraction. The paradox of the addition of a flow aid is that in order to disrupt the structure of the host powder it must itself structure strongly to the host particles.

Understanding the powder production process can also mean that the extremes of humidity and temperature have to be identified and that the structure might well be the result of several structuring mechanisms rather than one.

2.6 Tests for Effectiveness

Tests can be classified according to the degree of consolidation of the powder under test and will usually give a relative rather than an absolute measure of flow. Commonly great care is taken over the test procedure but the method of conditioning the host powder with the aid is ignored or dismissed as a minor detail. It is not a minor detail and should be carried out with care and understanding. The exposure of the test powder to variations in atmospheric humidity has a particularly insidious effect on powder conditioning. Typical flow tests used are by flow through orifices,[10] angle of repose,[8] bulk density determination,[17-19] shear and tensile failure.[20]

Another range of tests mimic actual process operations working with either pilot or full-scale equipment. Such tests can be sequenced in increasing consolidation as fluidisation tests,[21] storage stability[2,8,10] and tabletting.[4,9]

Table 1 *Some Common Flow Aids and their Properties*

Material	Manufacturer and Trade Name	Specify Gravity	Tapped Bulk Density (kg/m3)	Hausner Ratio	Particle Mean Size (μm)	Particle Form	B.E.T. Surface Area (m2/g)
Aluminium oxide	Degussa Corp Teterboro, N.J.	3.2	80		13		100
Phosphates Tricalcium phosphate:	Stauffer Chemical, Westport, Conn.						
	T.C.P.		320		80		
	Monsanto						
Polyethylene glycol	Carbon & Carbide Chem. New York, N.Y.						
	Union Carbide Co.						
	Carbowax						
	P.E.G.						
Polysaccharides Depolymerised starch:	Grain Processing Corp. Muscatine, Ia.						
	MALTRIN®		730	1.4			
Depolymerised cellulose:	F.M.C. Corp., Philadelphia, Pa						
	AVICEL®		304		20 - 90	microcrystalline	

Table 1 *Some Common Flow Aids and their Properties (Contd.)*

Material	Manufacturer and Trade Name	Specify Gravity	Tapped Bulk Density (kg/m3)	Hausner Ratio	Particle Mean Size (μm)	Particle Form	B.E.T. Surface Area (m2/g)
Silicates							
Silicondioxide (SiO$_2$)							
Precipitated:	J.M. Huber Havre de Grace, Md.						
	Zeofree 80®	2.0	144		14	amorphous powder	140
	Zeosyl 110S.D.®	2.0	224		40	spray dried powder	140
	Zeofree 5175B®	2.0	288		350	granular powder	180
	Zeothix 265®	2.0	95		4	micronised powder	250
	Degussa Corp., Dublin, Ohio						
	Sipernat 22S®	2.1	120		18	fluffy white powder	190
	Sipernat 50S®	2.1	99		8	fluffy white powder	450
	P.P.G. Ind. Pittsburgh						
	FLO-GARD FF®		130		10		180

Table 1 *Some Common Flow Aids and their Properties (Contd.)*

Material	Manufacturer and Trade Name	Specify Gravity	Tapped Bulk Density (kg/m3)	Hausner Ratio	Particle Mean Size (μ.m)	Particle Form	B.E.T. Surface Area (m2/g)
Silicates (Contd.)							
Silicondioxide (SiO₂)							
Gel:	Grace Davison Columbia, Md.						
	Syloid grades	2.1	112 - 465		9	micronised powder	320 - 675
	Millennium Speciality Chemicals Inc.,						
	Glidden Pigments, Baltimore, Md.						
	SiLCRON G-100®	2.1	112		3	micronised powder	275
	SiLCRON G-600®	2.1	176		4.7	micronised powder	325
	SiLCRON G-910®	2.1	400		9	micronised powder	700
Fumed:	Cabot Corpn, Tuscola, Il						
	Cab-O-Sil™ EH-5	2.2	40	16	7	light, fluffy	380
	Cab-O-Sil™ PTG	2.2	40	16	14	light, fluffy	200
	Cab-O-Sil™ M590	2.2	40	16	14	light, fluffy	200
	Cab-O-Sil™ L-90	2.2	50	16.5	24	light, fluffy	90
	Degussa AG Hanau Germany						
	AEROSIL 200®	2.2	50		12	fluffy	200
	AEROSIL R972®	2.2	50		16	fluffy	110
	AEROSIL 380®	2.2	50			fluffy	380

Table 1 *Some Common Flow Aids and their Properties (Contd.)*

Material	Manufacturer and Trade Name	Specify Gravity	Tapped Bulk Density (kg/m3)	Hausner Ratio	Particle Mean Size (μm)	Particle Form	B.E.T. Surface Area (m2/g)
Silicates (Contd.) Sodium Silico-aluminates:	JM Huber, Havre de Grace, Md						
	Zeolex 7®	2.1	272		8	powder	115
	Zeolex 201®	2.1	272		35	spray dried	75
Calcium silicate:	JM Huber, Havre de Grace, Md						
	HUBERSORB 250NF®	2.1			20	fine powder	120
	HUBERSORB 5121®	2.1	368		33	spray dried	50
Sodium lauryl sulphate:	Du Pont, Wilmington, Del Mallinckrodt Inc., St Louis, Mo. Flexichem, Swift, Chicago MEGRET Ltd., Knowsley, UK						
Stearates Calcium stearate [Ca(C$_{18}$H$_{35}$O$_2$)$_2$H$_2$O]	Mallinckrodt Inc., St Louis, Mo. Flexichem, Swift, Chicago. MEGRET Ltd, Knowsley		88	1.1	45	milled crystals	
Magnesium stearate: [Mg(C$_{18}$H$_{35}$O$_2$)$_2$H$_2$O]	Mallinckrodt Inc., St Louis, Mo. MEGRET Ltd., Knowsley, UK		88	1.1	45	milled crystals	
Zinc stearate: [Zn(C$_{18}$H$_{35}$O$_2$)$_2$H$_2$O]	Mallinckrodt Inc, St Louis, Mo. MEGRET Ltd., Knowsley, UK		120	1.1	75	crystalline	
			80	1.1	50	milled crystals	

2.7 Methods of Incorporation

Incorporation of a flow aid on an industrial scale faces the same conditioning problems as in the preparation of a powder for the testing of flow characteristics. A different conditioning philosophy must be developed for a 'soft' spreading aid such as talc and the smaller 'hard' aids such as fumed silicas. Much of the tradition of flow aid addition has built up on the assumption of a binary system comprising a single chemical host powder and the flow aid. Increasingly mixtures are multi-component and an aid addition could be in competition for active sites on bulk ingredients with an essential minor ingredient in the formulation. Almost certainly a sequential mixing procedure must be developed in this case. On the same theme of sequential mixing it may be that one of the ingredients in a multi-component mixture has flow problems that needs to be rectified by a pre-mix with an aid prior to a secondary mixing with the bulk ingredients. An agglomerative, active drug destined for mixing and subsequent direct compression tabletting would be just such an example. Sequential mixing is inconvenient but might be essential if structure and mixture quality are to be optimised.

2.8 Add-on Process Advantages

This is a growth area for the use of aids in that the aid will not only improve flow but will also add a desirable product quality. The additional benefit could be as a carrier of a liquid perfume or colourant, to cause tablet disintegration, to structurally reinforce a composite, to modify surface finishes, to aid incorporation into a liquid or to provide insulating properties. This is probably a high value added product capable of carrying the higher cost of a very specific flow aid or combination of flow aids.

3 FLOW AID SELECTION ADVISER

Figure 1 illustrates the network of information inputs required to make a logical selection of a flow aid rather than the traditional panic attempt to remedy an immediate crisis. An argument can be made for additional research effort under each of the eight headings but in the author's opinion the weakest knowledge base is in the sectors of 'Methods of Incorporation' of a flow aid and in identifying the 'Add-on Advantages' that a flow aid can bring to a product. If there are still limitations in the understanding of flow aid selection a major step forward has been made in recent years with the insights into the basic structuring forces applying within powders.

A possible general flow aid selection adviser is illustrated in Figures 2 and 3. It acts as a skeletal logic for a flow aid selection, testing and application procedure and can be adapted for specific company needs. In the procedure a bullet point marks a decision point for the user and is accompanied by an information section to guide the decision making process. Such a logical procedure helps to ensure that major process influences or advantages are not overlooked and attempts to optimise both the speed and the efficiency of the search for the optimum flow aid.

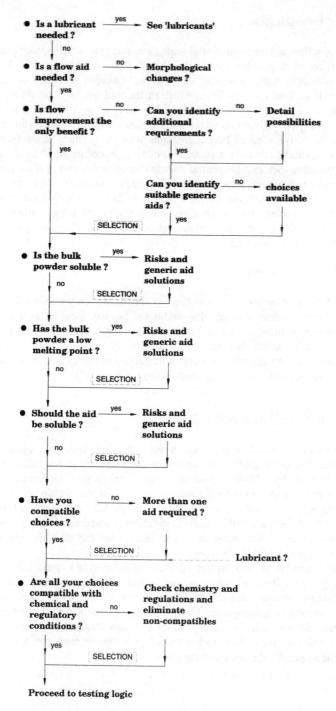

Figure 2   *Logic for a proposed selection procedure*

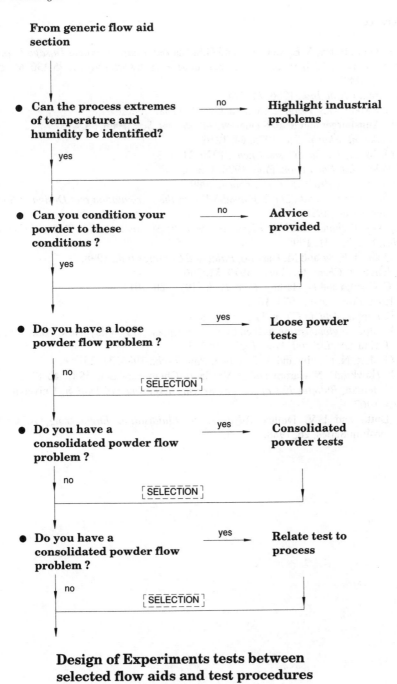

Design of Experiments tests between selected flow aids and test procedures

Figure 3 *Logic for a proposed testing procedure*

References

1. G. E. Peck, Baley, V.E. McCurdy and G.S. Banker, *Pharmaceutical Dosage Forms*, Edited by H. A. Lieberman, L. Lackman and J. B. Schwartz, Vol. 1, 75-130, Marcel Dekker, 1989.
2. N. Burak, *Chem. Ind.*, 1966, **21**, 844.
3. R. H. Striker, U.S. Patent 2,857,286, Oct 21, 1958.
4. L. L. Augsburger and R. F. Shangraw, *Jnl. Pharm. Sci.*, 1966, **55**, 418.
5. P. York, *Jnl. Pharm. Sci.*, 1975, **64**, 1216.
6. T. M. Jones, *Jnl. Soc. Cosm. Chem.*, 1970, **21**, 483.
7. M. Peleg, *Jnl. Food Proc. Eng.*, 1978, **1**, 303.
8. Grace, *Testing Procedures Handbook*, 1989.
9. Cabot, *Technical Data Sheet, Fumed Silica in the Formulation and Design of Solid Dosage Forms*, 1995.
10. Degussa, *Technical Bulletin Pigments, Synthetic Silicas as a Flow Aid and Carrier Substance*, No. 31, 1995.
11. D. Fluck, J. Fultz and M. Darsillo, *Paint and Coatings Ind.*, 1998.
12. H. Schubert, *Chem. Ing. Tech.*, 1979, **51**, 266.
13. M. C. Coelho and N. Harnby, *Pow. Tech.*, 1978, **20**, 201.
14. K. Hota, *Pow. Tech.*, 1974, **10**, 231.
15. C. E. Lapple, *Adv. in Chem. Eng.*, 1970, **8**, 1.
16. H. Rumpf, *Agglomeration*, Edited by W.A. Krepper, 379, Wiley, New York, 1962.
17. H. H. Hausner, *Intl. Jnl. Pow. Tech.*, 1967, **3**, 7.
18. D. Geldart, N. Harnby and A. C. Wong, *Pow. Tech.*, 1984, **37**, 25.
19. A. E. Hawkinds, N. Harnby and D. Vandame, Chem. Eng. Sci., 1987, **42**, 879.
20. A.W. Jenike, *Bulletin 123 of the Engineering Experimental Station*, University of Utah, 1967.
21. A. Dutta and L.V. Dullea, *Advances in Fluidisation Engineering*, A.I.Ch.E. Symposium Series, 26, 1988.

Measurement and Control

WHEAT FLOUR MILLING: A CASE STUDY IN PROCESSING OF PARTICULATE FOODS

Grant M. Campbell[1], Chaoying Fang[1], Philip J. Bunn[1], Andrew Gibson[2], Frank Thompson[2] and Arthur Haigh[2]

[1]Satake Centre for Grain Process Engineering, Department of Chemical Engineering, UMIST, Manchester, UK
[2]Microwave Group, Department of Electrical Engineering and Electronics, UMIST, Manchester, UK

1 INTRODUCTION

"It is probable that no two grains of wheat are exactly alike." Thomas Malthus, 1798. Thomas Malthus, in his *Essay on the Principle of Population* (1798), identified the inevitable balance that applies between food production and the population it can support, the power of population to increase being much greater than that of production. Cereals underpin human civilisation, with more than 50% of our global food intake coming from the three major cereals, maize, wheat and rice.[1] Undoubtedly improvements in cereal agriculture and productivity over the last several decades have contributed to the global population explosion; despite this, and the incredibly low market value of cereals, the quest is endless for ever higher yields and productivity. This is because cereals are, to the food economy, the equivalent of petrol in the total economy – cereal costs and availability impact on the cost-effectiveness of rest of the food manufacturing chain in a domino effect.

Despite the essential interaction between cereals and human society, the irony is that in the modern world, most of us are unaware of what the major cereals even look like. This is because cereals in their natural harvested state are somewhat unpalatable and indigestible, and must be processed into forms suitable for consumption. In the case of wheat, of which 600 million tonnes is produced annually around the world, the most common processing route is flour milling followed by some form of breadmaking. Both wheat and flour are particulate – wheat is granular, while flour is a powder, and the conversion of wheat into flour is a rare example of an entirely particulate food process.

Foods are almost never eaten in a powder form – sherbet is the single common exception. Powders lack texture and are dry and unappetising. Yet, particulate processes are common throughout the food industry, and many ingredients are supplied in particulate form, with the usual benefits of lowered transportation costs, shelf-stability and easy dispense, dissolution and reaction. An understanding of particulate behaviour and processing is thus essential in many food manufacturing industries. And wheat flour milling, being entirely particulate, is a good candidate for developing and demonstrating such an understanding. This paper describes current work aimed at modelling the evolution of the particle size distribution throughout flour milling, as a basis for design, optimisation and control of the process.

2 FLOUR MILLING: A COMPLEX PROCESS

Flour milling starts with the wheat kernel, which contains three main components:

- the **germ** (the baby plant);

- the floury **endosperm**, which supplies energy (stored as starch) and building materials (stored as protein) for the growing plant; and

- a protective coating of several botanically distinct layers known collectively as **bran**.

Milling of wheat into flour has two principle aims: to separate the bran from the endosperm; and to reduce the endosperm particles to a size sufficiently small to be rapidly hydrated on subsequent processing (*e.g.* during mixing of dough for breadmaking). In rice polishing, separation of bran from endosperm is achieved simply by abrading the outer bran layers away, leaving the intact endosperm particle. However, wheat (in common with barley, oats and rye) differs from rice in having an anatomical feature, the crease, in which the bran layers fold together into the centre of the grain.[2] While recent innovations such as the Satake PeriTec process[3,4] have introduced abrasion of wheat grains prior to milling, ultimately to separate wheat bran from endosperm, the kernel must be broken open. This is achieved using counter-rotating fluted rolls, which break the kernel open such that the bran particles tend to stay large while the endosperm particles are smaller, so that bran and endosperm can be separated based on size using plansifters. Using repeated roller milling and sifting in the "gradual reduction process" achieves highly efficient recovery of relatively pure endosperm material.

Figure 1 illustrates a typical flour milling process. The gradual reduction process uses primarily just two unit operations: size reduction by roller milling, and separation by sifting, linked together in complex configurations. Flour (endosperm material smaller than about 200 μm) is produced at each sifting stage. The process can be broadly divided into the Break system, which separates bran from endosperm, and the Reduction system, which reduces the size of endosperm particles. The Break system uses 250 mm diameter fluted rolls (with progressively finer flutes) operating at a differential of about 2.5, while the Reduction system uses smooth rolls operating at a lower differential of about 1.25.

To understand and model this complex process, three things are needed:

- models of particle breakage during roller milling (as a function of roller mill operation and wheat characteristics);

- models of particle separation by sifting (as a function of particle size distribution and sifter design and operation); and

- process models that link these together.

Such models, developed into interactive simulations, would be useful in process design and optimisation, process control, troubleshooting, and in training mill operators.

The process starts with First Break, which breaks open the wheat kernels. This broken material is then separated into five or more fractions, which are pneumatically conveyed to subsequent milling stages. Some flour is produced from First Break; this flour is of high purity and quality, with little starch damage, and may be sold as "patent flour".

Clearly, the particle size distribution produced at First Break determines the process flows and hence loadings and efficiencies throughout the rest of the mill. First Break receives an input of wheat that is invariably a mixture of several varieties (a "grist") and

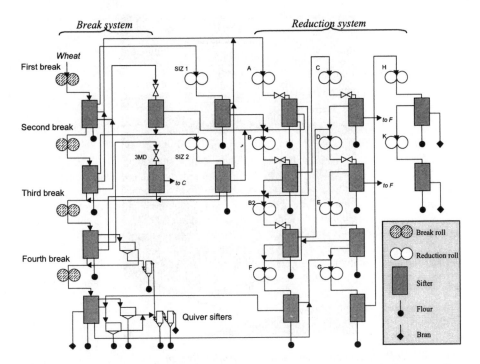

Figure 1 *Flowsheet for a typical flour mill with four break rolls (adapted from Owens[5])*

that is highly variable day-to-day and year-to-year. If, despite this variable feed, First Break could be continuously adjusted to deliver a consistent particle size distribution, this would simplify considerably the control of the rest of the process. Understanding breakage of wheat during First Break is therefore one of the critical control points for understanding, controlling and improving the flour milling process.

3 SINGLE KERNEL TESTING OF WHEAT QUALITY

Breakage of wheat grains depends on mill design and operation, and also on the physical and chemical properties of the kernels, such as size, shape, hardness, density and moisture content. Roller milling, in contrast to hammer or ball milling, is a once-through breakage operation, with no backmixing. Each kernel therefore breaks as a function of its own physico-chemical characteristics, independently of the surrounding kernels.[6] Breakage of a mixture of kernels depends on the distribution of these parameters in the sample. Kernel to kernel variation arises through genetic differences in the varieties gristed, and through environmental influences (location of the kernel on the wheat spike, variations across a field, across regions in a country, and across the globe.[7] As Malthus noted, no two kernels are identical.

In recognition of the importance of kernel-to-kernel variation, cereal quality testing is moving towards single kernel instruments, in which individual kernels are tested, and the distribution of parameters reported.[8,9] The most well developed commercial example of this is the Perten Single Kernel Characterisation System (SKCS), shown in Figure 2.[8,10-13]

Figure 2 *The Perten Single Kernel Characterisation System*

The SKCS samples kernels individually and weighs each, then crushes each kernel between a rotor and a crescent. A load cell on the crescent records the force profile during crushing, allowing the kernel diameter and a hardness index to be calculated. The electrical conductivity of each kernel is measured, from which the moisture content is calculated. The SKCS samples 300 kernels within 5 minutes, reporting the distributions of kernel mass, moisture, diameter and hardness in the sample.

One of the challenges with single kernel testing is to interpret distributions of measured parameters in terms of actual performance of the wheat on the mill. Until this link is made, single kernel testing is limited to serving as simply another quality control tool for acceptance testing of wheat at intake.

"Hardness" is an important wheat quality indicator, hard wheats generally having higher protein contents, milling more easily and being more suited to breadmaking. A mixture of hard and soft wheat varieties, for example, would be detected as a bimodal hardness distribution. Figure 3 illustrates the SKCS hardness distributions of Hereward, a typical UK hard wheat, and Consort, a soft wheat variety; both samples were from the 1999 UK harvest, grown in the Humberside region. Clearly the Hereward has a higher average hardness than the Consort, but there is a large overlap in kernel hardness between the varieties.

Hardness is a widely used but poorly defined wheat quality parameter, which can be defined loosely as "the way kernels break during milling, in terms of particle size and shape". Hard kernels break to give larger, more angular, progeny particles compared with soft wheats. It has been reported that hardness is correlated with kernel density.[14,15] Dobraszczyk *et al.*[15] suggested an explanation for the relationship between density and hardness, that the air spaces in soft wheats that lower the density might act as stress

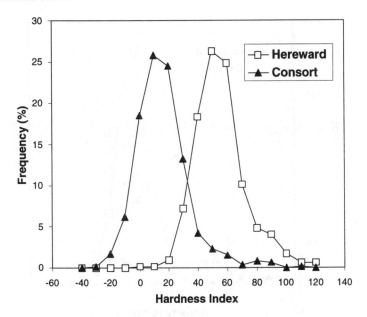

Figure 3 *Single kernel hardness distributions measured by the Perten SKCS, for Hereward (hard) and Consort (soft) wheats*

concentration regions, aiding kernel breakage and thus also lowering the hardness of the kernel.

Density is a fundamental, unambiguous parameter, and measurement of single kernel density may prove easier to interpret than hardness. Figure 4 shows the kernel density distributions of Hereward and Consort, measured using a double cup density measuring kit supplied by Precisa and described by Fang and Campbell.[16] Hereward has a greater average density and a smaller spread than the softer Consort, in agreement with the findings of Dobraszczyk *et al.*;[15] again there is a large overlap between the two varieties.

Figure 5 shows the average density *versus* hardness of 20 wheat varieties covering a range of hardness levels. The correlation is imperfect; however, if the samples are separated into those varieties accepted as hard, and those recognised as soft (of which there are only three in this set), the data suggest perhaps that separate correlations between hardness and density apply within the hard/soft classifications.

Work at the University of Manchester Institute of Science and Technology (UMIST) is aiming to develop a single kernel mass, moisture and density meter, based on the disturbance of a microwave resonant cavity when a kernel is placed in it; this allows calculation of mass and moisture, which is then combined with a novel single kernel volume meter to give density. Figure 6 shows the single kernel moisture measured by the microwave cavity as compared with that given by the SKCS, for wheat samples prepared to 10, 14, and 17% moisture contents (wet basis). This collaborative project at UMIST between the Microwave Group of the Department of Electrical Engineering and Electronics, and the Satake Centre for Grain Process Engineering, part of the Department of Chemical Engineering, funded by the EPSRC, aims to relate single kernel properties to wheat breakage during First Break roller milling.

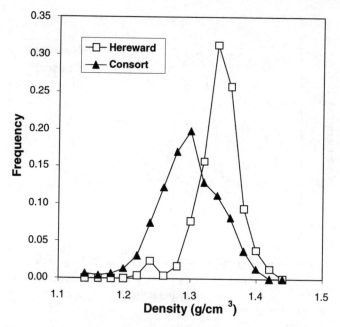

Figure 4 *Single kernel density distributions for Hereward (hard) and Consort (soft) wheats*

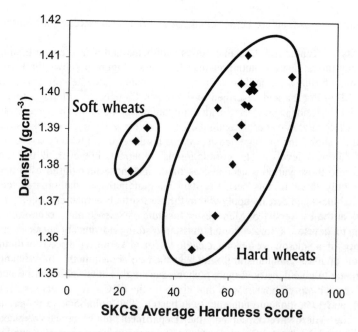

Figure 5 *Average single kernal density versus average SKCS hardness for 20 wheat varieties*

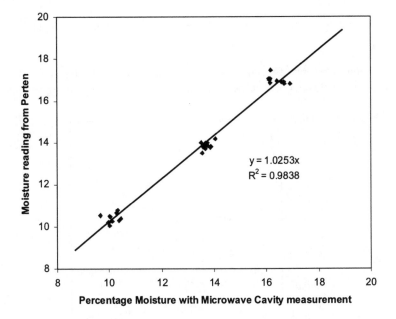

Figure 6 *Single kernel moisture measured by microwave cavity and by Perten SKCS*

4 BREAKAGE OF WHEAT DURING FIRST BREAK ROLLER MILLING

Campbell and Webb[17] and Campbell *et al.*[18] showed that wheat breakage during roller milling could be described mathematically by

$$\rho_2(x) \quad = \quad \int_{D=x}^{D=\infty} \rho(x,D)\,\rho_1(D)\,\mathrm{d}D \tag{1}$$

where $\rho_1(D)$ is the probability density function describing the size variation of the feed (the wheat), $\rho_2(x)$ is the probability density function describing the particle size distribution of the outlet material, and $\rho(x,D)$ is the breakage function, *i.e.* probability of creating an outlet particle of size x *given* an inlet particle of size D. Equation (1) is the breakage equation for roller milling. It assumes wheat kernels break independently during roller milling, as demonstrated by Bunn *et al.*[6] Knowing the breakage function for wheat kernels of size D, for a given variety, the breakage of a distribution of kernel sizes can be integrated to predict the output particle size distribution. A different breakage function is needed for each wheat variety; once established, these can be combined to predict breakage of mixtures.

These workers showed that breakage depended on the milling ratio, *i.e.* the ratio of roll gap, G, to kernel thickness, D (equivalent to SKCS diameter). The breakage function for First Break milling under a sharp-to-sharp roll disposition was found by milling narrow size fractions at different roll gaps, and could be described by an equation that was quadratic in G/D and linear in x over the range 250 μm $< x <$ 2000 μm:

$$\rho(x,D) = b_0 + 2c_0 x + \left(b_1 + 2c_1 x\right)\left(\frac{G}{D}\right) + \left(b_2 + 2c_2 x\right)\left(\frac{G}{D}\right)^2 \quad (2)$$

Figure 7 shows the form of this breakage function as measured for Hereward and Riband (another soft wheat variety), for different milling ratios. At a given milling ratio, the slope

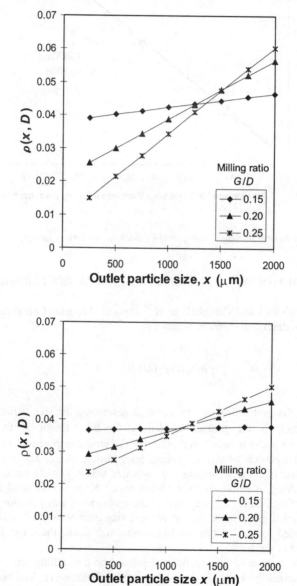

Figure 7 *Particle size distributions given by Equation (2) for various milling ratios, for (a) Hereward; and (b) Riband. From Campbell et al.*[18]

for Hereward (the hard wheat) was greater than for Riband, indicating a greater proportion of larger particles (the area under the line indicates the proportion of particles within a size range). This is to be expected for hard wheats, which under given milling conditions do not break to the same extent, resulting in proportionally more large particles (*i.e.* they are "harder" to break). As milling ratio increased (*i.e.* larger roll gap or smaller feed particle size), less breakage occurred and the slope of the line increased. Using this simple breakage function, Campbell *et al.*[18] showed that accurate predictions of breakage of whole wheat samples and of mixtures could be made.

5 EFFECT OF ROLL DISPOSITION ON WHEAT BREAKAGE

The above results were obtained for milling under a sharp-to-sharp (SS) disposition, as illustrated in Figure 8. Roller mills can also be operated under sharp-to-dull (SD), dull-to-sharp (DS) and dull-to-dull (DD) dispositions, also shown in Figure 8, resulting in different breakage patterns. To investigate the effect of disposition on breakage, samples

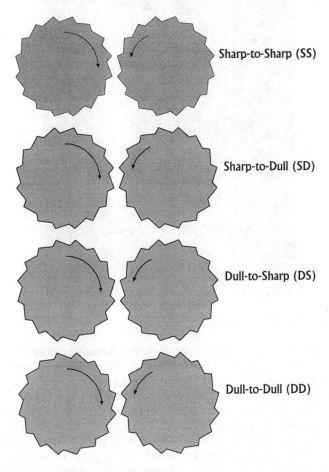

Figure 8 *Roller mill dispositions (not to scale)*

of Hereward and Consort were separated into narrow size fractions and milled at different roll gaps using the Satake STR-100 test roller mill, using the procedures described by Campbell *et al.*[18] Figure 9 shows the STR-100 test roller mill, which is fully variable in terms of roll fluting, speed, differential, gap and feed rate. First Break rolls, 250 mm in diameter with 10.5 flutes per inch were used under SS, SD, DS and DD configurations, with a differential of 2.7 against a fast roll speed of 600 rpm. Samples of 100 g were milled and collected for sieve analysis. Sieve analysis of the entire milled stock was performed on a Simon plansifter for 8 minutes using a sieve stack comprising wire mesh sieves of size: 2000, 1700, 1400, 1180, 850, 500, 212 µm, along with a bottom collecting pan. Breakage functions were constructed from the milling of these narrow size fractions, following the procedure of Campbell *et al.*[18] Equation (1), the breakage equation, was then used to predict the milling of whole wheat samples, and compared with experimental results at different roll gaps.

Figure 9 *The Satake STR-100 test roller mill*

Figures 10(a) & 10(b) show the particle size distributions obtained from milling whole wheat samples of Hereward and Consort at different roll gaps under SS, SD, DS and DD dispositions. Clearly, for both wheats, breakage under a dull-to-dull disposition is very different from that obtained using sharp-to-sharp milling. The latter gave results which could be broadly described by straight lines, the slope of which depended on the roll gap. Dull-to-dull, by contrast, gave breakage patterns that were more 'U'-shaped, with many more small particles, and many more large particles, and few in the middle size ranges. Decreasing the roll gap caused the 'U' to pivot, reducing the quantity of large particles and increasing the quantity of small particles, but having little effect on the middle range. SD and DS dispositions gave results intermediate between the SS and DD

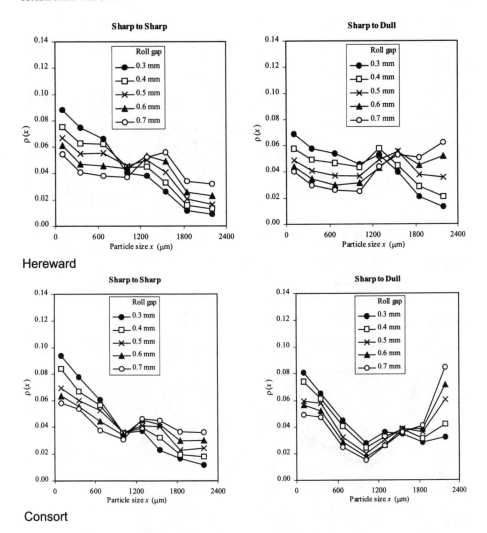

Figure 10(a) *Particle size distributions resulting from milling whole wheat samples of Hereward (hard) and Consort (soft) wheats at different roll gaps under SS and SD dispositions*

extremes. The transition from SS, through SD, DS to DD milling reflects the relative angle of approach of the flutes and their interactions with the kernels. First Break milling tends to be operated dull-to-dull, maximising the efficiency of separation of large branny particles and smaller endosperm particles. Figure 11 shows the particles obtained on breakage of Hereward and Consort under sharp-to-sharp and dull-to-dull dispositions. Dull-to-dull milling gave larger, flatter branny particles, while those from sharp-to-sharp were more angular.

Breakage functions were constructed by milling samples of narrow size fractions and different roll gaps. It was found that quadratic functions in x were appropriate to describe the breakage patterns for SD, DS and DD milling. To compare predictions using these

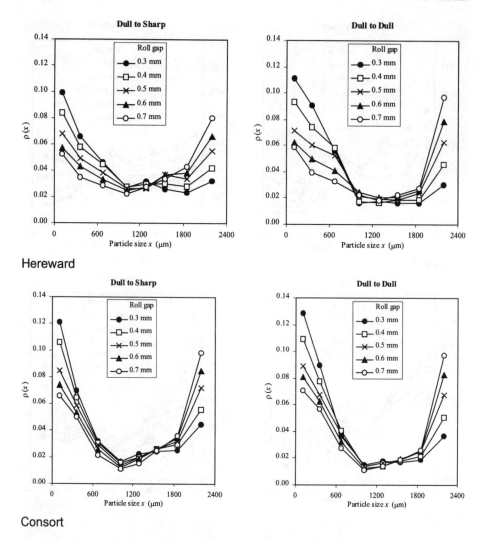

Figure 10(b) *Particle size distributions resulting from milling whole wheat samples of Hereward (hard) and Consort (soft) wheats at different roll gaps under DS and DD dispositions*

breakage functions with the experimental results, cumulative particle size distributions were plotted. Figures 12(a) & 12(b) show the cumulative particle size distributions predicted using the cumulative equivalent of Equation (1) (plotted as lines) compared with the experimental data (plotted as symbols), derived independently from milling whole wheat samples. Clearly the agreement is impressive, with R^2 values in excess of 0.99, indicating that this simple equation, with a simple breakage function form, is able to predict the entire particle size distribution in the range 250-2000 µm for any roll gap and any feed size distribution.

The breakage function, $\rho(x, D)$, describes the particle size distribution produced on breakage, *given* an initial particle size D, for a given wheat variety. As noted above, other

(a) Hereward, Sharp to Sharp

(b) Hereward, Dull to Dull

(c) Consort, Sharp to Sharp

(d) Consort, Dull to Dull

Figure 11 *Particles produced on First break roller milling of Hereward and Consort under sharp-to-sharp and dull-to-dull dispositions*

factors affecting wheat kernel breakage include hardness (h), moisture content (m), protein content (p) and density (ρ). For given operating conditions, the breakage function could be written more generally as $\rho(x \text{ given } G/D, h, m, p, \rho)$. Knowing the distribution of each of these factors, breakage could in principle be predicted for any wheat, without knowing the variety, or for a mixture of varieties, basing predictions entirely on

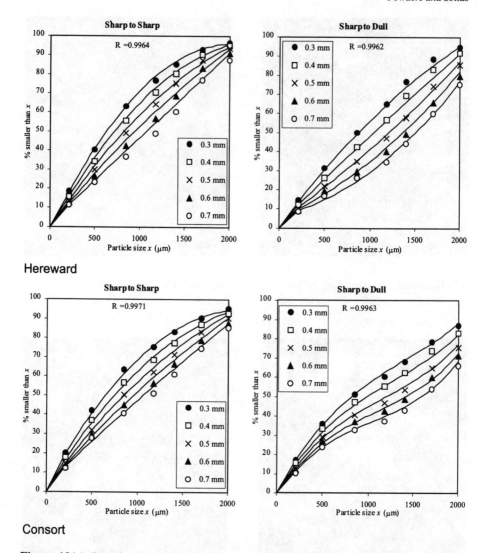

Figure 12(a) *Cumulative particle size distributions resulting from milling Hereward and Consort at different roll gaps under SS and SD dispositions: comparison of predictions using the breakage equation with experimental results*

measurement of single kernel characteristics. This provides a basis for linking the information from single kernel tests to actual milling performance. Work is ongoing at UMIST to extend the breakage function approach to incorporate single kernel moisture, hardness and density.

6 CONCLUSIONS

Flour milling is a rare example of an entirely particulate food process, which may be

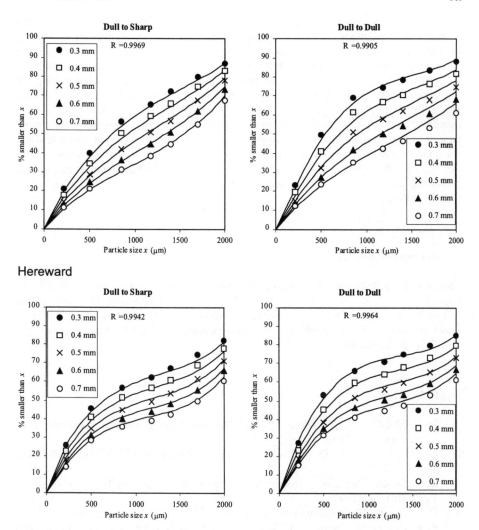

Hereward

Consort

Figure 12(b) *Cumulative particle size distributions resulting from milling Hereward and Consort at different roll gaps under SS, DS and DD dispositions: comparison of predictions using the breakage equation with experimental results*

viewed as "the evolution of the particle size distribution". In the gradual reduction process, floury endosperm is separated from bran with high efficiency using elaborate combinations of size reduction and separation operations. The process starts with First Break roller milling, in which breakage of the wheat is a complex interaction between mill design and operation and the physico-chemical characteristics of the wheat kernels. Wheat quality testing is moving towards measuring distributions of single kernel characteristics. Understanding and controlling breakage patterns during First Break is the key to interpreting single kernel tests and linking the information to milling performance.

The breakage equation approach to modelling roller milling relates the input and output size distributions over a roller milling operation, allowing a breakage function to be defined and quantified. Breakage for sharp-to-sharp milling can be described adequately, over a wide range of output particle sizes, denoted by x, using a function that is linear in x and quadratic with respect to the milling ratio. For dull-to-dull milling, the breakage function proves to be closer to a quadratic in x, with sharp-to-dull and dull-to-sharp dispositions in between. Dull-to-dull milling produces a clearer distinction, compared with other dispositions, between large branny particles and smaller endosperm particles. Predictions based on breakage functions, derived from milling of narrow size fractions of wheat, agree closely with data from milling whole wheat samples. This demonstrates the potential of the breakage equation approach, which could be extended to encompass distributions of other single kernel parameters, thus providing the link to milling performance.

Acknowledgements

This work is funded by the Engineering and Physical Sciences Research Council (EPSRC Grant GR/M49939). PJB is grateful to the EPSRC for an Engineering Doctorate award, in collaboration with RHM Technology. The assistance of the Satake Corporation of Japan in establishing the activities of the Satake Centre for Grain Process Engineering at UMIST is gratefully acknowledged.

References

1. P. C. Morris and J. H. Bryce, *Cereal Biotechnology*, Woodhead Publishing Ltd., Cambridge, UK, 2000.
2. N. L. Kent and A. D. Evers, *Kent's Technology of Cereals,* 4th Ed., Elsevier Science, Oxford, 1994.
3. R. S. Satake, 'Debranning process is new approach in wheat milling', *World Grain,* 1990, **8(b)**, 28, 30-32.
4. D. E. Forder, 'Flour milling process for the 21st Century' in *Cereals: Novel Uses and Processes*, G. M. Campbell, C. Webb, and S. L. McKee, (Eds), Plenum Press, New York, 257-264, 1997.
5. W. G. Owens, 'Optimisation and recycle grinding applied to flour mill processes', PhD Thesis, UMIST, Manchester, 2000.
6. P. J. Bunn, G. M. Campbell and S. C. W. Hook, 'Prediction of particle size distribution arising from First Break milling of wheat mixtures', *Proceedings of Process Engineering of Cereal Products Workshop*, Montpellier, France, 8th October 1999, published by ICC, Austria,1999.
7. M. B. Whitworth, , 'Heterogeneity in structure and grain composition', *Proceedings of Process Engineering of Cereal Products Workshop*, Montpellier, France, 8th October 1999, published by ICC, Austria, 1999.
8. C. Martin, R. Rousser and D. L. Brabec, 'Development of a single-kernel wheat characterisation system', *Trans ASAE*, 1993, **36**, 1399-1404.
9. A. D. Evers, 'New opportunities in wheat grading', 26th Nordic Cereal Congress, 1996.

10. R. Satumbaga, C. Martin, D. Eustace and C. W. Deyoe, 'Relationship of physical and milling properties of hard red winter wheat using the single kernel wheat characterisation system', *Assoc. Operative Millers Bull.,* 1995, January, 6487-6496.

11. C. S. Gaines, P. F. Finney, L. M. Fleege and L. C. Andrews, 'Predicting a hardness measurement using the Single-Kernel Characterisation System', *Cereal Chemistry,* 1996, **73**, 278-283.

12. B. G. Osborne, Z. Kotwal, A. B. Blakeney, L. O'Brien, S. Shah and T. Fearn, 'Application of the Single-Kernel Characterisation System to wheat receiving testing and quality prediction', *Cereal Chemistry,* 1997, **74**, 467-470.

13. C. Deyoe, Y. Chen, P. V. Reddy, P. McCluskey, J. Gwirtz and D. Eustace, 'Utilization of the SKWCS information in hard red winter wheat surveys for quality evaluations', *Association of Operative Millers Bulletin,* 1998, July, 7131-7140.

14. Y. Pomeranz and P. J. Mattern, 'Genotype and genotype x environmental interaction effects on hardness estimates in winter wheat', *Cereal Food World,* 1988, **33**, 371-374.

15. B. J. Dobraszczyk, M. B. Whitworth, J. F. V. Vincent and A. A. Khan, 'Single kernel wheat hardness and fracture properties in relation to density and the modelling of fracture in wheat endosperm', *J. Cereal Science* (in press).

16. C. Fang and G. M. Campbell, 'Effect of measurement method and moisture content on wheat kernel density measurement', *Trans. IChemE Part C, Food and Bioproducts Processing* (in press).

17. G. M. Campbell and C. Webb, 'On predicting roller milling performance: I. The Breakage Equation', *Powder Technology,* 2001, **115(3)**, 234-242.

18. G. M. Campbell, P. J. Bunn, C. Webb and S. C. W. Hook, 'On predicting roller milling performance: II. The Breakage Function', *Powder Technology,* 2001, **115(3)**, 243-255.

EXPERIMENTAL OBSERVATIONS OF POWDER CONSOLIDATION

A. de Ryck

École des Mines d'Albi-Carmaux
Route de Teillet
81013 Albi cedex 09
France

1 INTRODUCTION

1.1 Cohesion

Bulk solid handling remains a delicate issue in industry especially when powders are cohesive, i. e. when the interparticular forces are of the same order or greater than the weight. The origin of these forces are dipole interactions (van der Waals forces), electrostatic interactions if the particles are charged, capillarity if the particles are bound by a wetting liquid meniscus or these forces may raise from the formation of solid bridges between the grains. In this latter case, the mechanisms are for example sintering or re-crystallisation from the vapour phase or from a liquid film.

Figure 1 illustrates the competition between the two forces. The rupture occurs when the weight $P = \rho g S D$ exceeds the force of cohesion $F = cS$, so for:

$$D > D^*, \text{ where } D^* = \frac{c}{\rho g} \tag{1}$$

In these expressions, ρ is the apparent density of the powder, g the acceleration due to gravity, S the surface of rupture and D the thickness of the block which comes apart. c is

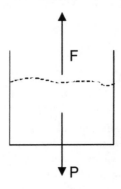

Figure 1 *Cohesion in particle beds*

called the *cohesion* of the powder, it has the dimension of a pressure and may be evaluated by counting the number of contacts on the failure plane. The force of cohesion is given by $F = N\kappa f$, where N is the number of particles on the surface and κ the number of contacts per grain. The interparticular force f may be expressed in the form: $f = \gamma a$, where γ is a surface energy (of order 10^{-3} mN/m)[1] and where a is a length scaling the size of the contact area. a ranges from a nanometric scale to the size d of the grains. The surface is $S = N d^2 / (1 - \varepsilon)$, where ε the porosity of the granular medium. The ratio of these formulae leads to this scaling:[2]

$$c \sim \frac{\gamma a}{d^2} \kappa (1 - \varepsilon)$$

(2)

Equation (1) demonstrates that the cohesion c is a key parameter for the scaling of silos and all installations where powders are manipulated since it gives the minimum size allowing flow. From Equation (2), it can be noted that this cohesion depends on the size of the particles: the smaller they are, the more cohesive they behave. The second dependency is that the cohesion increases with the compaction of the powder: there is less porosity (ε), more contacts for each grains (κ) and more robust (a). This effect is taken into account in the Jenike method for silo design when determining the flow function.[3]

1.2 Ageing

There are also temporal effects. For most materials a small compaction during time under normal stresses can be observed. The main explanation is that the contacts involve by plasticity, sintering, but also small rearrangements of small amounts of particles, leading to more efficient contacts. For vibration induced compaction, the second mechanism should be preponderant. Experiments on creep[4] or vibro-compaction[5] of dry particles show that they lead to a weak logarithmic increase of the cohesivity with time.

An another way for consolidation is humidity adsorption on wettable powders. For relative humidity H below saturation (H < 1), a capillary condensation should occur, limited by the size given by the Kelvin law. This leads to the formation of menisci at a nanometric scale bridging two contacting particles. This is a metastable, thermally activated process, yielding a force between particles, which increases logarithmically with time:[6]

$$f \sim \frac{\gamma a}{\ln H} \ln \frac{t}{\tau_0}$$

(3)

Here τ_0 is a microscopic time, γ is the surface tension of water and a a distance which takes into account the size of the grains and their nano-rugosity.

Up to this point, we have a description of the physical origin and practical implication of the cohesion of a powder and a description of two logarithmic ageing mechanisms. One due to grain rearrangements and the other one due to the evolution of the contact forces by capillary condensation. In the second part, we propose to describe some experimental observations of powder consolidation to compare with these theoretical descriptions.

2 EXPERIMENTAL OBSERVATIONS

All the experiments have been performed using an annular shear tester sketched in Figure 2 and described in detail elsewhere.[7] Three powders with different lumpiness will be successively described: a non-cohesive silicagel powder, dry and humid salt and a fungicide for potatoes.

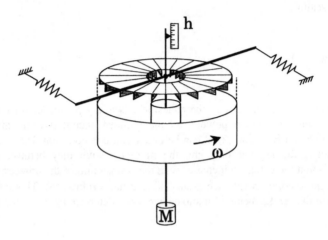

Figure 2 *Annular shear tester*

2.1 Fine Non-cohesive Powder

We used some silicagel 60, ground and sieved between 40 and 63 µm. Due to its microporosity (the mean diameter of the pores is 60 Å), this powder well absorbs humidity and appeared to be quite insensitive to changes in relative humidity. It then exemplifies the case of a powder which only experiences van-der-Waals interactions between the grains.

We performed holding experiments as shown in Figure 3 where the friction coefficient μ, ratio of the shear and normal stresses, is plotted versus time. The powder is sheared till a plateau is obtained for the shear force. Then at point A, the shear tester is stopped and the powder left to relax during the holding time T. Finally, at point B, the shear is resumed and we observe a peak in the force to restart the flow.

This peak increases logarithmically with the holding time, as can be observed in Figure 4, which shows the maximum of friction coefficient μ_{max} versus T.

For this material, this consolidation process is associated with a small compaction. In Figure 5, both the coefficient of friction and dilatancy of the sample during hold are plotted versus time. We observe that the compaction is also logarithmic with time leading to a power-law relationship between μ_{max} and the packing down dh during hold, with an exponent of about 1.5 ± 0.5. But the way the sample compacts remains unclear: do the grains rearrange (a collective effect) or do we observe the plasticity of the contacts (local effect).

Finally, we may note that this ageing is very small indeed. In order to double the force

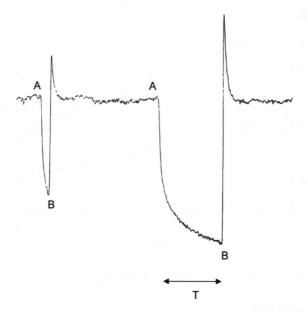

Figure 3 *Friction coefficient μ versus time for two successive holding experiments. A: the motor is stopped; B: the shear is resumed*

necessary to restart the flow, one needs to wait more than the accepted age of the Universe. This silicagel, though showing a measurable consolidation, does not lump in practice.

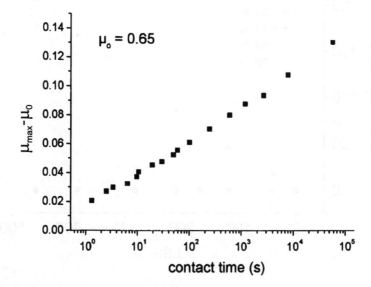

Figure 4 *Peak of friction after a holding time*

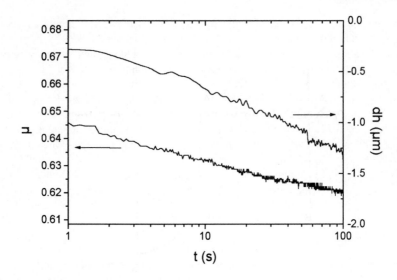

Figure 5 *Temporal variation of μ and dilatancy h during the hold*

2.2 Dry and Humid Salt

In order to observe the consolidation due to moisture adsorption, we have studied the shear response of an assembly of NaCl crystallites. Their size is around 0.5 mm. Before

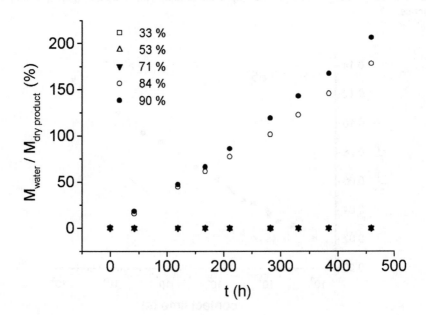

Figure 6 *Mass of water adsorbed versus time for different relative humidity*

the experiments, the salt was left during two weeks in a chamber with controlled relative humidity. At laboratory temperature, the equilibrium value for the relative humidity H with salt is 75 %. If the air is more humid, we observe a constant water adsorption (Figure 6) and a thick salt saturated water film covers the surface of the NaCl grains.

For the shear and hold experiments, the salt is put inside a cell which cannot be totally closed in order to avoid friction between the moving and immobile parts of the device. This can be inconvenient for the interpretation of the long duration experiments since water vapour may escape from the gap left between the cover and the walls of the cell. To quantify that problem, we have performed some experiments for which the ageing took place in the controlled humidity chamber (under the same normal stress but without any shear stress). The evolution then observed is similar to the one observed in the shear cell with the same stresses conditions.

The observation of the kinetics of consolidation (Figure 7) shows two distinct behaviours. For dry salts, we observe a weak logarithmic ageing as for the silicagel. This dry behaviour is identical for all the relative humidities under the equilibrium value of 75%. This result differs from what is observed for glass beads. Bocquet et al[6] observed a condensation kinetics below the dew point. Above the equilibrium relative humidity, the shear behaviour of the salt changes. For short times, we still observe a logarithmic increase of the cohesion of the powder, but more rapid. The extrapolation of this short time behaviour leads to a doubling of the force necessary to restart the flow in only 10 years. But the kinetics observed is even more rapid than that. After a few hours, the cohesion sharply increases and the doubling finally occurs after 10 hours.

This acceleration in the kinetics takes place whatever the macroscopic quantity of water adsorbed since the experiments performed for two weeks and 5 weeks matured salt in a 95 % relative humidity give the same results. Its origin needs to be clarified by doing more controlled experiments. The first point to check is the role of the interstitial gas. In

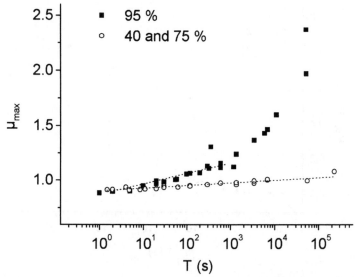

Figure 7 *Maximum friction coefficient versus the holding time for humid and dry salt*

our procedure, the cell is filled in ambient air so that the gas has a humidity of 40-50 %. It may be sufficient to dry the powder sample for long experiments. The second point to check is the role of the grain dissolution which may accelerate the sintering. We plan to perform experiments with insoluble particles and a non-volatil liquid to separate the capillary effects from mass transfer effect between the phases.

Finally, it may be emphasised that, even if the kinetics is not totally understood, the first minutes of observation are sufficient to distinguish the dry and humid behaviour. For this material, it is then possible to propose a short time control test.

2.3 A Fungicide Powder

The last example is a fungicide powder which presents lumping. The more there remains some traces of one of the reactants (up to a few ‰), the more difficult it is to handle the powder. This reactant is known to sublimate at ambient temperature, which may explain this increasing difficulty to manipulate this powder after storage.

We have tested 3 samples of this fungicide, labelled 1 to 3, containing an increasing quantity of this reactant (from 0 to 3 ‰) using the holding experiments. The result, given in Figure 8, does not follow the qualitative classification given by the users. Furthermore, there is here no possibility of distinguishing between the three samples by looking at the beginning of the kinetics: up to 20 minutes, there is no difference in the way they consolidate. The reasons for these discrepancies between the qualitative observations and quantitative measurements of consolidation are surely that the experiments do not reproduce the same history of lumping. In particular, in an industrial environment, this powder may be exposed to cycles of sublimation-recrystallisation, which is not the case in our experiments. A remaining way of investigation then is to understand how the vapour pressure cycles may enhance the kinetics of consolidation.

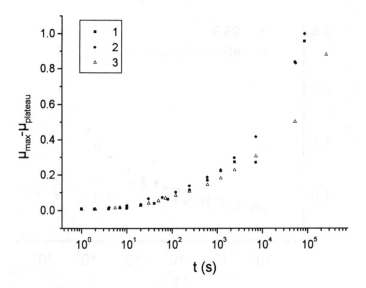

Figure 8 *Consolidation of three samples of fungicide*

3 CONCLUSIONS

We have presented several experimental observations of consolidation kinetics. For dry particles like free-flowing silicagel or salt, we have observed a weak logarithmic increase of the force necessary to restart the flow.

In the presence of mass transfer (humidity adsorption and dissolution or sublimation-recrystallisation, the kinetics is faster than logarithmic. Further investigation to understand that phenomenon and more experiments with an accurate control of the vapour pressure are needed.

This last point is a necessity when trying to fix a lumping problem in industry. One needs to understand these vapour pressure fluctuations during process and storage in order to understand how the mass transfers operate in time in order to devise adequate remedies to this problem.

References

1. K. L. Johnson, K. Kendall and A. D. Roberts, *Proc. Roy. Soc. A*, 1971, **324**, 301.
2. H. Rumpf, *Chem. Ing. Tech.*, 1970, **42**, 538.
3. A. W. Jenike, *Bull. Univ. Utah Eng. Exp. Sta.*, 1964, **123**.
4. C. Marone, *Annu. Rev. Earth Planet. Sci.*, 1998, **26**, 643.
5. E. Ben-Naim, J. B. Knight, E. R. Nowak, H. M. Jaeger and S. R. Nagel, *Physica D*, 1998, **123**, 380.
6. L. Bocquet, E. Charlaix, S. Ciliberto and J. Crassous, *Nature*, 1998, **396**, 735.
7. M. Lubert and A. de Ryck, *Phys. Rev. E*, in press.

PROCESS AND· POWDER HANDLING IMPROVEMENTS RESULTING FROM PARTICLE EMISSION MEASUREMENT AND POWDER MASS FLOW MONITORING USING ELECTRODYNAMIC TECHNOLOGY: - A CASE STUDY

William Averdieck

PCME Ltd
Clearview Building
Edison Road
St Ives
Cambs PE27 3GH

1 INTRODUCTION

Operators of pneumatic transfer systems and spray drying systems are looking more and more to improve the efficiency of the operation, by minimising the potential for product loss from the dust collection system and by detecting changes in powder flow sufficiently early to avoid costly downtimes caused by blockage. This paper describes how new electrodynamic type particle emission and powder flow instruments have been successfully used in a Powder Milk facility to address this long-standing problem.

In 1995 two DT-770 particle emission instruments were installed in the emission stack of a milk spray dryer to monitor product loss from the cyclone. After two years of successful results on the stack, the process operator and PCME jointly evaluated a particle mass flow monitor, which used the same instrument technology, but with a non- intrusive ring sensor, in a pneumatic transfer system on the same spray dryer. These trials gave the process operator sufficient confidence in the performance of the instrument to install the system on a further four lines. This paper discusses the operation of both the emission and flow instruments and how they have performed in the process. It also refers to the process benefits arising from these measurements being made reliably. The spray dryer is located in North West England at a powdered milk facility and the instruments evaluated were manufactured by PCME.

2 PROCESS DESCRIPTION

Liquid milk is sprayed into a chamber through which hot air is being circulated. Water in the milk evaporates· producing a fine particle of powdered milk and steam. The resulting mixture of steam, air and powder flows from the chamber and then into a cyclone where the powder is separated from the gas steam. The powdered milk is thrown to the outside of the cyclone through centrifugal action where it is collected and then flows out of the bottom of the cyclone to the agglomeration system. The agglomeration system involves re-injecting the powdered milk fines back into the spray drier at a number of radial positions via a lean phase pneumatic transfer system. The re-entrainment in the spray dryer results in the particles agglomerating to produce a larger desired particle size. However in the process of transferring the milk fines in the pneumatic transfer system flow disruptions can occur in the 50mm pipes due to the high humidity conditions, which results in complete blockage of the pipes and closure of the spray drier. The process operator had, therefore, great interest in finding a powder flow meter, which could reliably monitor and detect changes in flow conditions in the pneumatic transfer system.

In the same spray drying process, air and steam separated from the powder in the cyclone pass out of the top of the cyclone and are emitted to atmosphere through an 800mm duct. The cyclone by definition is not 100% efficient, so small quantities of powdered milk are carried in the air/steam stream to atmosphere. In addition process upsets (i.e. build-ups within cyclone) which reduce the cyclone's efficiency temporarily, can result in larger quantities of particulate being lost up the stack. This is undesirable due to concerns regarding product loss, environmental emissions and increased clean up costs. There was, therefore, an additional requirement to install a reliable particle emission instrument in the stack, which could measure and monitor the amount of milk powder being lost/discharged to atmosphere. This is a difficult monitoring requirement due to the high levels of water vapor in the stream, which can cause interference and instrument fouling.

A schematic of the process and the location of measurement points are given in Figure 1.

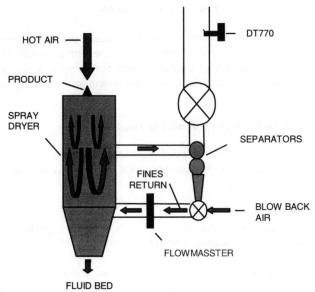

Figure 1 *Process schematic*

3 ELECTRODYNAMIC TECHNOLOGY

The instruments installed to satisfy both the powder flow and particle emission requirements use a form of electrodynamic technology. While the sensor configuration used is different for each type of instrument, the principal of operation of both types of electrodynamic type instruments is the measurement of the alternating current which results from charge induction as distributions of charged particles pass a grounded sensor element (Figure 2). Under the right conditions this alternating (ac) current can be related directly to particulate concentration. In a system used to monitor particle emissions the

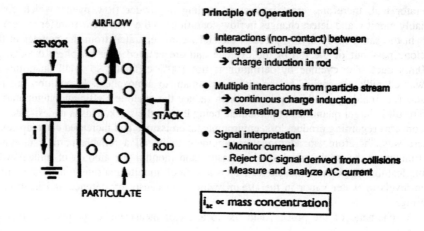

Figure 2 *Principle of operation of electrodynamic sensor*

sensing element is a rod with a conductive core. This sensing probe is installed across part of the stack in a position to be exposed to a representative profile of the particle emissions. In the case of the powder flow instrument the sensor is a ring through which the flow passes. This permits a representative measurement without being intrusive in the flow.

3.1 Electrodynamic Technology for Emission Measurement

The interaction between sensor element and particle is similar for both types of sensor. For the rod type emission sensor, particles produce a charge movement in the probe as a result of two kinds of interaction:

1. Direct collision.
2. Charge induction (charge on the particle repels charge in the probe as it passes). This is the dominant interaction with small (<100 microns) particles.

The variation in the particle distribution results in an alternating component of the current, and since the particle distribution follows a Poisson distribution; the amount of variation is directly proportional to the number of particles. Hence for a constant particle size and charge distribution, the ac current is proportional to the dust concentration,[1] provided the chosen bandwidth for the ac signal is not susceptible to mechanical or electrical interference. In making the electrodynamic measurement, a dc or triboelectric current is also generated by the particles colliding with the sensor rod, however this is specifically filtered out by the electronics, since this is significantly influenced by velocity and probe contamination.

Independent bodies such as Warren Spring Laboratory and TUV[2] have validated the proportional and repeatable relationship between the ac current and dust concentration in emission applications. For example the results from the linearity tests conducted by TUV on a dust collector application are shown in Figure 3.

As is the case with other types of particulate emission monitoring techniques, there are limitations on the usage and capability of electrodynamic technology. First the

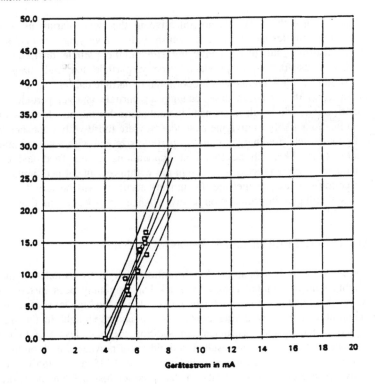

Geräterstrom in mA

Figure 3 *TUV linearity test results*

instrument's calibration against dust concentration will shift with changes in the following process conditions:

- Changing particle size distribution.
- Changing particle type.
- Changing pre-charge on particle.

In a spray dryer application the obvious concern is the effect of changing humidity conditions on the pre-charge on the particle. However field experience has shown that once moisture levels at elevated temperatures (>100°C) reach above 10% the effects of changing moisture levels on the ac signal are not significant. This means in practice that when the spray drier is on and moisture levels are varying there is no effect on calibration. The condition in which the calibration may change is when particles are present with very low levels of moisture. This occurs at start up and shut down of the process.

Secondly the validity of ac measurement techniques is dependent on the assumption that the particle distribution follows a statistically predictable relationship. This assumption is maintained in the large majority of emission applications in which the dust particles are well distributed within the air stream and the flow is relatively constant on a second to second basis.

3.2 Electrodynamic Technology for Powder Flow Monitoring

The ring sensor of powder flow monitor derives its signal from induction as particles pass through the sensing element. Unlike the rod sensor there is very little signal coming from particle collision, since the sensor is not intrusive in the flow. Variations in particle distributions as the particles pass through the sensor induce a varying electrical signature pattern. In a uniform dilute phase conveying stream, the particle distribution again follows a Poisson distribution, so provided velocity, particle size and charge remain constant this signal has an ac component proportional to particle concentration.

Of course as in the emission case, changes in particle charge, particle size and statistical distributions will result in changes in calibration for the powder flow monitor. Therefore applications likely to give the best accuracy are those with a constant particle type (this effects particle pre-charge), constant particle size and with a non-pulsing flow. These conditions are likely to be found in applications in the food and chemical processing industries where processes by design run within certain defined limits. In cases where these conditions are not met the instruments accuracy will be less, however the output signal will still give information on relative changes in process conditions.

4 RESULTS WITH PARTICLE EMISSION MONITOR

Opacity type instruments that measure the absorption of light by particles as they pass through a light beam transmitted across the stack had originally been installed on the stack. However these had never satisfactorily worked due to the high level of water vapor resulting inherent in the spray drying process which absorbed light in the frequency spectrums used in the opacity instrument. In addition the high levels of humidity in combination with particle emissions resulted in build-up of powdered milk on the optical surfaces necessitating regular instrument maintenance. In 1995, a decision was taken to find alternative measurement instruments so that product losses from the spray dryer could be better understood and avoided.

A number of measurement techniques were initially considered, however the DT-770 (Figure 4) was selected for evaluation for the following reasons:

1. The output of the instrument was stated to be linear even in the presence of high water vapor levels.
2. The use of an insulated probe was possible, hence avoiding the problems of insulator shorting and contamination, which are common with triboelectric and electrodynamic sensors when used in conductive (high humidity) environments.
3. The instrument could be calibrated in mg/m^3 by isokinetic sampling and the calibration was repeatable.
4. The instrument included multichannel capability and integral datalogging useful for internal reporting.
5. The instrument had automatic zero and span checks to ensure valid measurement.

The instrument was supplied with a 600mm intrusive sensor rod sufficient for representative measurement across the 800mm stack. The sensor was specified for temperatures up to 250°C providing a safety margin over the 200°C temperatures normally experienced in the stack. The sensor was mounted via a 1½ inch BSP coupling in a straight section of duct which gave greater than 2 diameters of straight before the sensor.

The instrument was initially tested for a 60-day period to ensure fit for purpose. This evaluation was facilitated by the integral datalogger that enabled data to be easily compared and tracked. The following characteristics were established:

Figure 4 *DT-770 instrument*

1) The output of the instrument was dynamic and tracked known changes in process condition (i.e. increases in production, plant shut downs, and simulated dust emissions with injected dust).
2) The output of the instrument was stable and repeatable for the same process conditions over the full duration of the test.
3) The instrument required no maintenance.
4) The instrument provided useful data to the operator to control the emissions.

Subsequent to the trial the system was upgraded to include a sensor for a second stack. The instruments were then calibrated in mg/m^3 by correlating the average output from the instruments to the results of an isokinetic sample and reprogramming the calculated scaling factor into the instrument.

Since the initial trial period the instruments have continued to exhibit the same practical and robust performance first shown in the trial. In addition repeated isokinetic samples have shown the instrument calibration to be relatively constant inspite of the high humidity levels. Admittedly it has not been possible or practical to perform calibrations over a full range of process conditions, however the user's confidence in the instrument to provide on-line information on changing dust levels has further increased. The 4-20mA output from the instrument is tracked in the control room so that action can be taken to tune the process when increases in particle emissions are detected. The datalogger is used to report average emissions for internal environmental emissions reporting.

5 RESULTS WITH POWDER FLOW MONITOR

The return fines from the cyclone had never been satisfactorily monitored in the pneumatic transfer system in the process. Historically, ultrasonic and acoustic type flow monitors[3] had been tested in the application, however, they had not provided satisfactory results. Therefore, site glasses which required regular inspection were being used to approximately balance the flow in the various lines and visually confirm flow was occurring. Blockages were seldom detected with the site glasses since this would have required continuous surveillance.

In late 1996, after receiving a proposal from PCME to evaluate a new instrument, two Flowmasster flow meters with non-intrusive ring sensor (Figure 5) were installed in the return fines lines to continuously monitor the reduction in flow rate of milk powder and capachino powder and by doing so to prevent blockages.

Figure 5 *Powder flow monitor installed in spray dryer*

The ring sensors comprised the same internal diameter as the pipe bore (50mm). Having a flush non-intrusive internal finish they were also hygienic which was critically important for this food application. The sensor spool piece was mounted in the line with dairy pipe IDF fittings. The temperature in the line is ambient.

On initial power-up the system provided an instantaneous reading of relative powder flow rate. This signal tracked known changes in powder conveying rate.

The sensors were subsequently calibrated by diverting the flow from the line after the sensors and weighing the collected sample. The instrument was calibrated in Kg/s assuming a constant velocity by correlating the instrument output to the weighed sample. A trial unit was closely monitored and checked against weigh scale samples. A difference was noted between the response of the instrument to capachino powder and powdered milk. It is thought that this was due to the different dielectric coefficients of the two materials, which influenced the amount of pre-charges that the powders obtained before passage through the sensor ring. The effect was shown to be repeatable and could be compensated for by using different calibration factors. The calibration of the Flowmasster with a particular type of powder was also shown to be relatively constant (+/- 10%) which was perfectly adequate for the intended process control use.

An extensive 12-month trial established that the units operated without the need for maintenance and were able to prevent problems in the spray drier by warning of changes in the particle flow before blockages started. In addition due to the relatively constant process conditions it was additionally possible to calibrate the instruments to measure the amount of powder in the agglomeration system, which permits further process optimisation.

6 SUMMARY OF PROCESS BENEFITS

The following process benefits have been obtained as a result of using the information from the emission monitor and powder flow monitor:

Emission Sensor:
- Reduced product loss from process
- Emission incidents minimised
- Reduced clean up costs associated with incidents
- Credible data generated on environmental compliance
- Reduced emissions

Flow sensor:
- Better flow balancing
- Early detection of flow disruptions
- Optimisation of process
- Blockages avoided
- Payback of 3 months on instrument cost due to reduced downtime and disruptions

References

1. K. Smolders and J. Baeyens, University of Leuven, *Continuous Monitoring of Particulate Emissions by means of a Tribo-electric (electrodynamic) Probe*, Powder Handling & Processing, 1997, **9 (2)** April/June.
2. *TA Luft Certification of DT-770*, TUV Essen.
3. Y. Yan, *Mass flow measurement of Bulk Solids in Pneumatic Pipelines*, Measurement Science Technology ,1996, 1687 –1706.

Subject Index

Tri calcium phosphate (continued)
  as flow aid (continued)
    manufacturers of, 85
    properties of, 85
TUV, 122-123
  and linearity of dust concentration tests,
    122-123

Unconfined failure tester, 5
Uni-axial tester, 7
Union Carbide Co., 85
University of Manchester Institute of
    Science and Technology (UMIST),
    99
University of Newcastle, 5
University of New South Wales, 5
University of Wallongong, 5
Utah University, 4

Valve
  cone, *see* Cone valve
  rotary, 30
  split butterfly, 44, 47
Vent
  panels
    positioning, 27
    sizing, 27
Venting
  and design of vent cover, 20
  on a duct, 22
  of explosion, 19-21
  flameless, devices, 21
  indoors, 21
  and vent duct, 20
  and vent opening, 20
Vertical shear
  cell, 7
    and static failure, 7
  tests, 8
Vibration, and wall friction, 7
Vibratory feeder, *see* Feeder, vibratory
Void ratio, 4
  and bulk density, 4

Wall friction
  and design, 6
  importance of, 9
  instruments to measure, 7

Wall friction (continued)
  and silo wall pressures, 4
  tests, 8
    for bulk storage operations, 8
    for new solids handling, 8
  effect of vibration on, 7
Wall pressures
  of silos, 4
Warren Springs Laboratories, 5-6, 122
Weigh feeder, *see* Feeder, weigh
Welding and cutting
  as ignition source, 18
Wheat
  breakage
    and first break, 101-103, 105, 107
    mathematical description, 101-103
    and roll disposition, 103-107
    and single kernel properties, 98-101,
      108-109
  Consort, 98-99, 104, 107-109
    and particle size distribution, 104,
      107-109
  hard, 98
  Hereward, 98-99, 102, 104, 107-109
    and breakage function, 102-103, 107
    and milling ratios, 102-103
    and particle size distribution, 102-104,
      107-109
  kernel density, 98
    of single kernel, 99-100
  kernel hardness distributions, 99-100
    for Consort, 100
    for Hereward, 100
  quality of, 97
    using Perten Single Kernel
      Characterisation System (SKCS),
      97-101
    single kernel testing, 97-101, 109
  Riband, 102
    and breakage function, 102-103
    and milling ratios, 102-103
    and particle size distributions, 102-103
  soft, 98
  varieties, and SKCS hardness, 100
Wheat flour milling, 95-110, *see also* Flour
    milling
  using fluted rolls, 96
  using plansifters, 96

RETURN TO: **CHEMISTRY LIBRARY**

100 Hildebrand Hall • 642-3753

LOAN PERIOD	1	2		3
4		1-MONTH USE 5		6

ALL BOOKS MAY BE RECALLED AFTER 7 DAYS.
Renewable by telephone.

DUE AS STAMPED BELOW.

NON-CIRCULATING UNTIL: 4/23/63 3 PM		

FORM NO. DD 10
3M   3-00

UNIVERSITY OF CALIFORNIA, BERKELEY
Berkeley, California 94720–6000